JEFF BRANUM

972-8031

Structural and Foundation Failures

Structural and Foundation Failures:

A Casebook for Architects, Engineers, and Lawyers

by
Barry B. LePatner, Esq., &
Sidney M. Johnson, P. E.

McGraw-Hill Book Company

New York   St. Louis   San Francisco   Auckland
Bogotá   Hamburg   Johannesburg   London   Madrid
Mexico   Montreal   New Delhi   Panama   Paris
São Paulo   Singapore   Sydney   Tokyo   Toronto

Library of Congress Cataloging in Publication Data

LePatner, Barry B.
  Structural and foundation failures.

  Includes index.
    1. Structural failures—Case studies.
  1. Building failures—Case studies.   I. Johnson,
Sidney M. II. Title.
TA656.L46   624.1   81-5986
ISBN 0–07–032584-7   AACR2

1234567890   KPKP   8987654321

The editors for this book were Joan Zseleczky and Iris Cohen,
the designer was Mark E. Safran, and the production supervisor was
Teresa F. Leaden. It was set in Baskerville by the Kingsport Press.

ISBN 0-07-032584-7

Printed and bound by the Kingsport Press.

Contents

Preface

Each year, construction totaling billions of dollars is completed in the United States. It is a compliment to the design profession that relatively few design-related problems arise. Such problems do occur, however, and, starting in the mid-1950s, claims against architects and engineers began increasing in dramatic proportions. This increase in claims continues and has engendered extensive interest and concern to both the legal and design professions.

The design professionals' primary concern in this regard is to avoid such claims. It is in their interest to learn the lessons which such claims teach, in order to avoid repeating them in their own practices. Unfortunately, this is far easier said than done because of a natural reluctance of professionals to publicize their difficulties. As a result, a designer has the opportunity to learn of problems, and the lessons which might derive from the consideration of these problems, only from those few cases which attain sufficient notoriety to be published in the media or in trade journals, or where the designer, or friends and associates, are personally involved. Moreover, the quality of such information tends to be poor—accounts in the media are often incomplete and, at times, inaccurate—word-of-mouth accounts are often little more than rumor.

The complete repository of accurate information concerning actual design and construction problems is the files of the lawyers who have (on either side) handled the claims which have grown out of these problems. However, this data is not readily accessible. The authors believe that the public interest would be served, subject to the obvious need to protect the names and reputations of the parties involved, if some of the data from these files were published. This is the primary purpose of this book.

The lawyer also will find an interest in these claims. His interest rests with the proper handling of such matters and the pragmatics of representing the design professional. Barry B. LePatner, an attorney practicing in New York City, has devoted a substantial portion of his professional career to the representation of architects and engineers and their professional liability insurers. Mr. LePatner has handled design professional litigation throughout the country, encompasing the full range of the topics covered in this book. These topics include theoretical aspects of representation as well as practical examples taken from actual cases.

The engineering analysis which follows each chapter has been prepared by Sidney M. Johnson, P.E., a principal in the consulting firm of Berkowitz-

Johnson, Consulting Engineers. Presentation of the authors' experience and a summary of pertinent theory is the second function of the book.

The book is set up so that each case is covered in the following format:

Part 1. Narrative of events

Part 2. Discussion of relevant technical problems

Part 3. Discussion of legal and contractual points

In this fashion, those having a general interest can read Part 1 of each case, those involved with the law can continue with Part 3, and those interested only in the technical problems of design and construction can concentrate on Part 2.

The cases described are actual cases, simplified and idealized for ease of understanding and have been selected to illustrate general principles and more common problems, rather than unique, once-in-a-million, occurrences. For this reason, no case citations or legal references have been included.

The various arguments in each case are presented. The authors do not endorse them as applicable to every situation. In most cases, there is no one correct approach to handling a design defect claim, largely because many such matters are so complex that there is no simple or single solution. Indeed, in several instances, the authors find items of technical appraisal and of legal adjudication with which they do not agree. In some instances these doubts may be expressed. More often they are not. No apology is made for this. The purpose of the book is to present facts and every attempt has been made to present sufficient, in-depth information to provide meaningful, specific lessons of practical application. It is only in the technical commentaries that the authors presume to venture opinion and to guide the reader to formulate a conclusion.

The authors wish to express their appreciation to Donald R. L'Abbate whose able assistance and guidance was invaluable during the formative stages of the book.

The authors are also deeply indebted to Virginia Sheppard, Colleen McGill, and Debra Caleo, who assisted in preparing the manuscript and helped nurse the many revisions that were required over the months of its preparation.

Barry B. LePatner
Sidney M. Johnson

Structural and Foundation Failures

Introduction

As architects and engineers are aware, there has been a dramatic increase in the number of claims brought against their professions in recent years. In 1960 the claim frequency per 100 architectural-engineering firms was 12.5. By 1969 the frequency per 100 firms had increased to 20, and in 1974 the frequency was just under 25 claims per 100 firms. The latest projected figures provided by various insurers of design professionals show a frequency increase to nearly 35 claims per 100. The increased frequency of claims has become a burden to the design profession, detracting from production time and entailing great expense, both direct and indirect.

As a result of this claim frequency, the design professional now finds the need to spend increasingly more time meeting with attorneys, in conferences and in court appearances. Such use of time, of course, is nonproductive from an income point of view. In addition, the design professional finds that malpractice insurance rates have dramatically, even prohibitively, increased during this period.

With the advent of this explosion of claims, the design profession has become increasingly aware of the legal pitfalls it faces in today's practice. The legal profession likewise has recognized that to properly represent the design profession, it must be fully versed in all of the intricacies of the design process, as well as in the latest decisions affecting the construction industry. The legal practitioner must recognize that in representing an architect or engineer, one is entering into a specialized field of law requiring a working knowledge of the practices and relationships of the design profession. Also, one must be prepared to cope with the intricacies of construction contract documents and the vagaries of construction disputes, as well as with the nomenclature common to these disputes. The legal practitioner who represents design professionals in all aspects of their practice must be prepared to review and study proposed contracts, continuously update those contracts so that they conform to the latest statutory and decisional law, and know how to use that decisional law to upgrade the client's position in ongoing projects. In short, today's attorney who represents architects and engineers must serve not only a legal function but an educational one as well.

Architects and engineers in practice today must ensure that their attorneys are fully conversant with this area of expertise and possess full knowledge of the design professional's activities. Architects and engineers, for their part, must be prepared to constantly hone their business and technical skills in order to improve their practice. Their attorneys can be helpful in this regard. In turn, the architect or engineer should teach the legal professional the intricacies of the construction industry. It is through a proper marriage of the two professions that the aims of the design profession are best served. The architect or engineer must be prepared to devote time to both preventing claims and working cooperatively in the defense of claims when they arise.

Construction disputes are costly creatures. What may start out as a simple problem can often encircle and encompass the unwary designer and attorney in a web of complications involving numerous parties, legal decisions, and an exposure far beyond that originally envisioned. The analysis of a claim itself is time-consuming and expensive. Voluminous documents which are developed by virtue of the construction process must be reviewed. Plans and specifications likewise must be analyzed in detail.

Every project has its own history. It is only when the attorney, after viewing the project with the perspective of hindsight, analyzes all the facts, reviews the relationships among the parties, and fully understands the documents, personalities, and problems attendant to a specific claim that he or she can hope to prepare a reasoned analysis in order to make a recommendation to a client. There is no shortcut to representing the design professional in a construction dispute.

For the architect or engineer, the primary concern is not to protect oneself from a claim but to avoid the claim in the first instance. The architect and engineer is interested in a "zero defects" practice and, generally, is greatly interested in learning the lessons of experience—preferably someone else's experience, a less painful alternative to one's own. The following cases cite the facts, as actually presented by the several adversary parties, and attempt to draw forth from these facts technical lessons of value to designers and attorneys alike.

These lessons are of considerable practical importance. In each case described, the architect or engineer involved was a capable, licensed professional. It should be apparent to the reader that the problems cited in the cases often are caused by situations not covered in the standard educational texts. Indeed, some of the lessons are likely to be revealing even to seasoned experts.

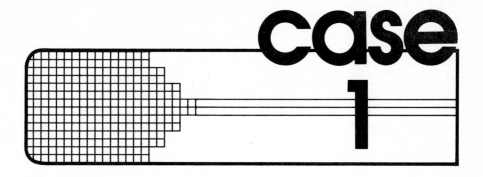

case 1

TYPE OF FACILITY	Parking garage
TYPE OF PROBLEM	Cracking of posttensioned, multistory, rigid frame

Significant Factors

A. Shrinkage problems
B. Creep strains due to posttensioning
C. Thermal effects
D. Construction sequence
E. Overestimation of dead load

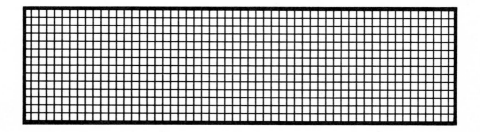

NARRATIVE

This project involved a multilevel parking garage for passenger vehicles. Because of restrictions on the deeding of the property, it was necessary to provide, in addition to parking stalls, recreational amenities as part of the structure. The architect produced an imaginative design which involved utilizing a portion of each parking level as a recreational plaza. The recreational plaza supported an earth fill and landscaping, including pavements, plantings, and other features. Because the site was on a hill and each floor was level, not sloped, access to each of the levels was provided by street entrances, thereby eliminating the need for ramps.

The structural system developed for the building was a posttensioned, concrete, rigid frame. Shortly after construction commenced, the structural engineer, during one of his site visits, noted an unusual cracking pattern around one of the beam-to-column connections. Having had little experience with posttensioned structures, unable to account for the cracking, and fearing that it might be more than cosmetic, he wisely called for help. A review of his plans by an independent consultant revealed that a potentially hazardous situation existed.

At the time of discovery (construction had not proceeded beyond the first level), the state-owner was notified of the condition. Unfortunately, this event aggravated a previously existing conflict between the architect, the owner, and the contractor which had already resulted in at least a 6-month delay in construction of the project.

While remedial plans were being prepared by an independent consultant retained by the structural engineer, the contractor discontinued work, despite efforts by the architect to have the contractor continue work in relation to other available areas. In addition, the architect, believing that the services he was being called upon to perform were exceeding the scope of his original contract, demanded additional compensation. The owner was quickly depleting the appropriation for this project and was not anxious to go to the legislature for additional funding.

The matter soon developed to the point that the project stood close to abandonment, with damages increasing in geometric progression. In an effort to prevent or minimize further delays, details of the remedial work were produced piecemeal by the independent consultants. Although the contractor had in his possession sufficient drawings to recommence work, he still refused, stating that work would not begin again until he had *all* the remedial plans. The contractor refused to order additional reinforcing material and posttensioning tendons which would be required under the revised design. Consequently, when the contractor submitted to the owner a proposed change order and time extension on its contract, the architect refused to approve it. Pressed by the state and threatened with termination, the architect provided his own estimates of compensation due the contractor and recommended that the contractor be granted no extension on the contract time.

As each of the parties became more recalcitrant, a void was created in the management of the project. None of the parties was willing to make any decision with respect to the scheduling and and phasing of the work. The contractor adopted a posture of relying on total guidance from the architect and the architect refused to intrude upon the province of the contractor, namely, the means and methods of construction and construction scheduling. The owner demanded that the situation be remedied immediately but without making any constructive suggestions or providing any guidance.

Clearly, the technical problem (the design deficiency) was but the tip of the iceberg—and readily solved. The greater problem was to get the parties working again, together, to finish the job. As will be described under Legal Analysis, and as the result of long negotiation and conciliation, lines of communication were reestablished between the parties, the managerial void was filled, and the work completed. Upon completion, the parties negotiated their differences and a settlement, short of litigation, was achieved.

TECHNICAL ANALYSIS

Diagrammatic representation of this structure is presented in Figure 1-1.

Shortly after the first deck was poured, cracks were noted of the type shown in Figure 1-2. These cracks occurred at the intersection of the post-

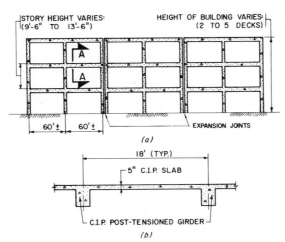

FIG. 1-1 Diagrammatic representation of multilevel parking garage. (*a*) Elevation. (*b*) Section A-A (typical deck).

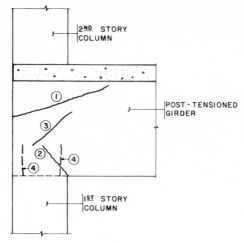

FIG. 1-2

tensioned girders with the exterior columns. The cause of the type 2 cracks was determined to be a combination of excessive forces and moments acting at the joints between the girders and columns due to shortening of the girders resulting from shrinkage of the concrete and to elastic and creep shortening resulting from posttensioning. Temperature effects also contributed to this shortening. Cracks of type 1 were determined to be a bearing type of failure due to compressive stresses occurring under the anchorage plates, coupled with inadequate reinforcement in the ends of the girders. The cause of the type 3 cracks was determined to be related to the bursting effect of the posttensioning forces. In general, cracks of types 1 and 2 appeared in all beams. Cracks of type 3 occurred at random locations.

Initially, corrective measures were taken consisting of additional reinforcement in the ends of the girders and changes to the bearing plates for the posttensioning cables. These measures did *not* adequately eliminate the cracking problem. Accordingly, the detail at the intersection of the girders and the exterior columns was altered to provide a slip joint, as shown diagrammatically in Figure 1-3. This modification further reduced the cracking, although it introduced new, lesser cracks of type 4. Additional column reinforcement was added as shown in Figure 1-4.

The cause of the type 1 cracks can be seen by reference to Figure 1-5. These cracks emanated from the bottom of the anchorage plate for the top posttensioning cable.

The cause of the type 2 cracks is indicated in Figure 1-6. The shortening of the girders results in shears and flexural distortions (moments), as indicated. Insofar as the girders are concerned, the effects are cumulative as

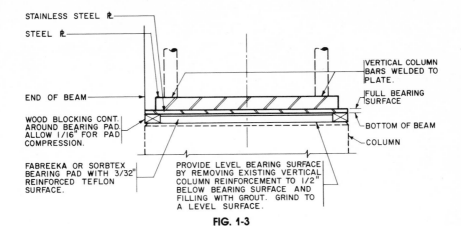

STAINLESS STEEL ℞

STEEL ℞

END OF BEAM

WOOD BLOCKING CONT. AROUND BEARING PAD. ALLOW 1/16" FOR PAD COMPRESSION.

FABREEKA OR SORBTEX BEARING PAD WITH 3/32" REINFORCED TEFLON SURFACE.

VERTICAL COLUMN BARS WELDED TO PLATE.

FULL BEARING SURFACE

BOTTOM OF BEAM

COLUMN

PROVIDE LEVEL BEARING SURFACE BY REMOVING EXISTING VERTICAL COLUMN REINFORCEMENT TO 1/2" BELOW BEARING SURFACE AND FILLING WITH GROUT. GRIND TO A LEVEL SURFACE.

FIG. 1-3

each succeeding level is put on the building. The causes of the shortening of the girders, as briefly noted above, are the following:

1. Shrinkage of the concrete

2. Elastic and creep strains due to posttensioning stresses

3. Temperature

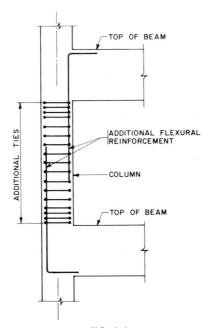

TOP OF BEAM

ADDITIONAL TIES

ADDITIONAL FLEXURAL REINFORCEMENT

COLUMN

TOP OF BEAM

FIG. 1-4

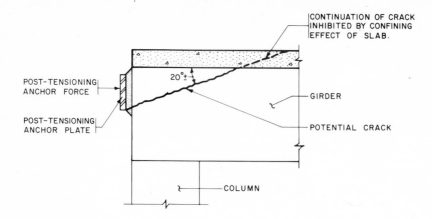

This is a common cause of distress in concrete rigid frames, particularly where the columns are stiff and exercise considerable lateral and flexural restraint on the girders. A dramatic example of the problem occurs in the design of bridge piers, where the pier columns, because they support heavy loads, tend to be short and massive. It is customary to relieve the distortions in bridge piers by specifying a pouring sequence in which at least the setting shrinkage is minimized (see Figure 1-7). Adaptation of this principle in building construction is not widely done, but should be.

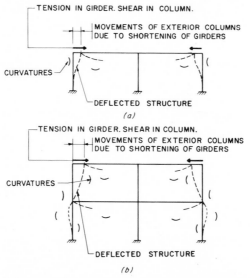

FIG. 1-6 Cause of type 2 cracks. (*a*) Distortions occurring due to shrinkage and stressing of first deck. (*b*) Additional distortions occurring due to shrinkage and stressing of second deck.

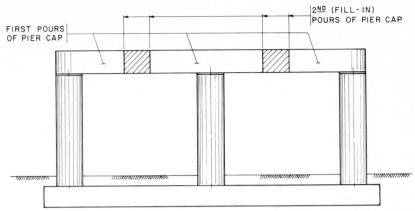

FIG. 1-7 Pouring sequence for minimizing setting shrinkage in bridge piers.

In the case being discussed, one of the points of the designer's defense was that he expected the structure to be poured in a vertical sequence (presumably one line of columns at a time), whereas the actual pouring was of large, horizontal areas. Points to note include the following:

1. Shrinkage of concrete comprises two components—setting and drying. A controlled pouring sequence can largely eliminate the effects of the setting shrinkage, but the drying shrinkage continues for substantial periods beyond the time which normally can be accommodated in the pouring schedule.

2. Lap splices should be provided in the reinforcement of the area of "fill-in" pours, i.e., permitting slip to occur due to shrinkage of adjacent pours.

3. Shrinkage can be reduced by using a lower water-cement ratio, using types I and II cement instead of type III, using standard weight instead of lightweight aggregate (standard weight concrete, in general, is better for posttensioned work than lightweight concrete), and chilling the aggregate and mixing water to lower the temperature of the concrete during placement.

The mechanism causing the type 3 cracks can be seen by reference to Figure 1-8. Such cracks are often noted in posttensioned beams and, generally, are not considered to be cause for concern about the structure. Palliative measures consist of increasing the thickness of the stressing (bearing) plate and adding end reinforcement (stirrups).

The cause of the type 4 cracks was variously determined to be (1) shrinkage of the concrete in the girders, resulting in drag on the slip joint assembly (these cracks occurred at the locations of the vertical column steel), or (2) bearing failure caused by the tendency of the posttensioning forces

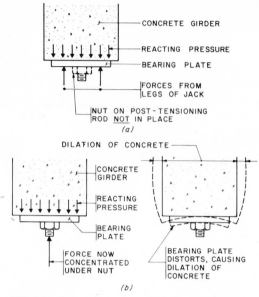

FIG. 1-8 Cause of type 3 cracks. (*a*) Step 1 of posttensioning process. (*b*) Step 2 of posttensioning process.

to lift the girders upward, causing the weight of the girders to be transferred to a limited bearing area.

A second problem which developed on this project was that the structural design of the posttensioned girders was started at an early stage of the development of the architectural design—at which time locations of proposed planting areas on the decks had not been established. The structural engineer assumed a uniform load, of conservative value, to represent the weight of the soil and boxes which would constitute the planting areas. In the final design, the planted areas were of limited size and variously located. Accordingly, the actual loads were *less* than those assumed by the structural designer, and differently distributed. For a prestressed design, this lack of imposed dead load and the different distribution thereof changes the final stress levels, since the assumed dead load stresses which offset the posttensioning stresses do not exist.

Comments

Other Causes of Cracking

Some of the potential causes of the observed cracking which initially were considered, but rejected, are of interest as a checklist for consideration in analyzing such problems.

1. Settlement of formwork

2. Excessive construction loads

3. Inadequate lap length in splicing reinforcement

4. Improper drape of tendons

5. Overstressing of tendons (checked by reference to stressing logs)

6. Internal honeycomb in the area of the distress due to the maze of reinforcing steel in this area

Use of x-ray techniques to check for items 3 and 4, although not actually used, was considered and might have application under other circumstances.

*Effects of Secondary and Parasitic Stresses
on Strength of Structures*

It is interesting to speculate as to whether, considering the inevitable costs and delays, it was essential to modify the design to eliminate the strains caused by the shortening of the girders. Sufficient details are not available to permit the necessary calculations. However, in principle, once cracking of the girders and/or of the exterior columns occurred so that equivalent (plastic) hinges were formed, the restraints causing the distress were totally or partially relieved. Figure 1-9 shows the effects of formations of plastic hinges. Flexural distortions of structure are relieved, and stability of structure is provided by the center column, which now must provide all required

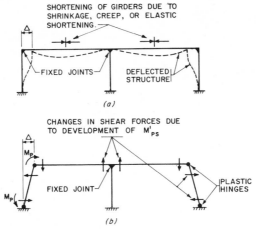

FIG. 1-9 (*a*) Conditions prior to formation of plastic hinges. (*b*) Conditions after formation of plastic hinges.

resistance to lateral loads (or sidesway). This is generally true of secondary and parasitic stresses, and the occurrence of such stresses often does not significantly affect the *ultimate* strength of structure, provided the following conditions are met:

1. Buckling is prevented, both of members and, in the case of metal structures, local buckling in areas of plastic hinges.
2. Plastic hinges can develop without causing something to "snap." In concrete design this is accomplished by limiting the reinforcement ratio to 0.75 P_B. In mild steel structures, it is an inherent result of the ductility (yield) of the material.
3. Shear stresses induced by the changed distribution of loads are not excessive.
4. There is sufficient redundancy that the development of the plastic hinges necessary to relieve the secondary stresses does not create a collapse mechanism.

The authors have had several occasions to use this principle in the evaluation of the adequacy of an existing structure, of the need to underpin structures which have been subject to differential settlement, and of the need to unload (see Figure 1-10) structures which were to be modified to carry increased loads (by adding stories to a building or by strengthening a floor). Some examples are as follows:

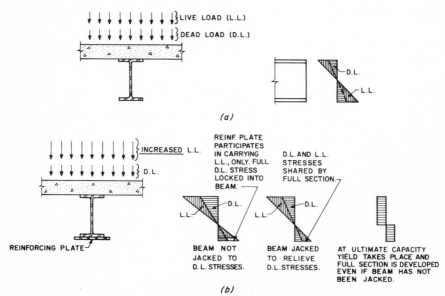

FIG. 1-10 Stress distribution in a strengthened beam. (*a*) Existing construction and stresses in beam. (*b*) Strengthened construction and stresses in beam.

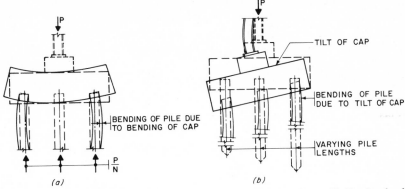

FIG. 1-11 Sources of deformation of pile cap. (*a*) Spanning of cap. (*b*) Varying load-displacement characteristics of piles.

1. In a conventional pile foundation, because of distortion of the pile cap (Figure 1-11) and variable final driving resistance, pile loads cannot be equal. It is conventional to design the pile caps as though they were equal, because at ultimate capacity, wherein the soil supporting the piles yields, the pile loads will be substantially equal, provided that the pile and the pile cap are strong enough that they do not fracture before the ground yields.

2. Figure 1-12*a* represents an arrangement for drydocking a ship. The sufficiency of this arrangement depends on crushing (yield) of the soft-

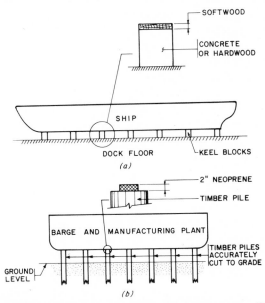

FIG. 1-12 (*a*) Conventional drydocking arrangement. (*b*) Adaptation used to eliminate the drydock.

wood to equalize the loads on the blocks. The authors have used an arrangement similar to this for docking a barge-mounted manufacturing plant on piling (Figure 1-12*b*).

3. Renovation of Yankee Stadium in New York City had to contend with the locked-in stresses incident to up to 6 in of differential settlement. The matter was handled by limiting L/r of all compression members to 40 and adding sufficient excess strength in all connections to assure that plastic yield of the members would occur before buckling and before the connections would fail.

LEGAL ANALYSIS

This case is illustrative of the problems a design professional can encounter by using techniques or procedures with which he or she is not fully conversant. Fortunately, the designer in this case was conscientious, and upon the discovery of the unusual cracking pattern for which he had no explanation, he sought assistance.

Early recognition of a construction problem is often the key to amicable resolution. What is necessary is an objective view of the situation, with designers neither acting as their own judge and jury, condemning themselves out of a sense of guilt, nor deluding themselves with the belief that the problem will remedy itself. Seeking the aid of counsel at an early stage provides a psychological advantage. In effect, the designer has engaged a partner to share the problem and to provide the objectivity which may be necessary for the ultimate identification and rectification of the problem. In this case, although the consequences of disclosure were harsh, had the problem not been recognized and disclosed and had the structure been built according to the original plans, the consequences might have been even more serious.

The facts in this matter exemplify a number of principles applicable to construction problems and construction litigation. First, projects with rigid budget restrictions (such as this one, where funding was through government appropriation) are more susceptible to delay damage claims. When problems are encountered on such projects and additional funds are required to rectify the problem, the source of those funds must be established before the problem can be rectified. Thus, it is necessary to establish liability before the project has been completed. Under such circumstances, the parties often react as adversaries and cooperation can diminish to the point of nonexistence. Such attitudes are not conducive to successful completion of the project; thus, the problem becomes amplified in scope. On projects where the owner is willing to fund and authorize change orders, corrective measures with a reservation of rights will enable the parties to act more objectively and be more cooperative. The questions of liability can then await completion of the project.

This case also illustrates that a negative approach to a construction problem will foster feelings of doubt and uncertainty. Admitting a problem but professing ignorance of its nature or denying outright responsibility fosters uncertainty and doubt in the mind of the owner. The designer is more likely to find a successful conclusion to a problem by approaching the owner with a positive and affirmative attitude in conjunction with a proposed plan of action. In this case, immediate consultation with an independent expert and the creation of a remedial plan enhanced the designer's position in the eyes of the owner. The owner was rightly impressed that the designer was concerned with the problem and was doing everything possible to minimize both delay and damage to the owner. The professionalism which the designer displayed in this matter helped maintain the client's confidence despite the design deficiency.

Notwithstanding the foregoing positive events, the project did come dangerously close to abandonment. It is noted, for the sake of the lesson to be learned here, that the design problem was the culmination of a series of events which had caused strained relations between the parties prior to its discovery. With the discovery of the design problem, there came the involvement of counsel, which was beneficial to the further handling of the project. Objective counsel for all parties were more readily agreeable to compromise in order to achieve completion of the project. All counsel involved were successful in dissuading their respective clients from unreasonable positions and were instrumental in preparing the matter for ultimate resolution.

The lawyer, in the role of counselor, advises the client regarding the relative strengths and weaknesses of the position the client wishes to maintain. A lawyer objectively views the position taken by the client and counsels the client in regard to the value or danger involved in holding such a position. Through private consultation the client is often made to realize that a position of complete vindication is not possible. If the client is willing to accept counseling by the attorney, the attorney is often able to effect a reasonable compromise through negotiation.

The attorney's function as advocate is to advance the position most advantageous to the client. However, the attorney will remain flexible and the ultimate position taken will represent a balance between the weakness and strength of the client's case.

The attorney's third role in a case such as this is to protect the client in confrontations with other parties. The designer in a situation such as the one presented would naturally be nervous and might not be the person best able to make a presentation concerning the remedial work. The lawyer, accustomed to such situations, is more likely to present an effective and positive presentation. The involved parties, when left to their own devices, will often return to arguing relative positions and mire themselves in details rather than in the means of solving the dispute. This, in turn, leads to unnecessary bickering and accomplishes little, if anything. Lawyers, however, will often not resort to arguing questions of fact or liability but will seek to find areas of agreement upon which goals can be established and achieved.

Lessons to Be Learned

1. Designers should be vigilant to possible problems throughout the project.

2. If potential problems should arise which bring any doubts to the designers' minds, they should seek outside counsel and assistance. Under no circumstances should problems be ignored nor overreactions occur.

3. Designers should exercise greater care and skill when using new techniques or using techniques with which they have had little or no experience.

4. Involving experienced construction litigation lawyers early in the process can be beneficial to designers and can be instrumental in amicable resolution of the problems.

5. Lawyers in these circumstances act as counsel and advocate on behalf of designer clients.

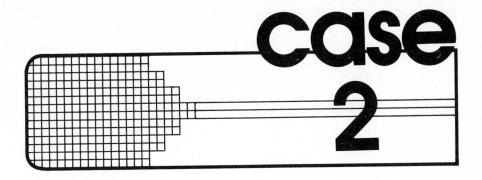

case
2

TYPE OF FACILITY   Bridge

TYPE OF PROBLEM   Cracking and distortion
of welded girders

Significant Factors   A.   Welding technique
B.   Methods for correcting distortion due to welding
C.   Effects of residual stresses

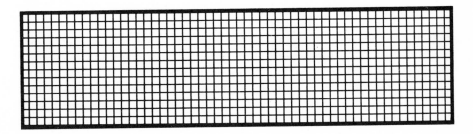

NARRATIVE

Engineering services are not limited to the preparation of plans and specifications. Often, they include on-site inspection during construction. Judgments required of the inspecting field forces may be highly subjective, and such decisions often become the subject of vigorous debate, leaving possible exposure for legal liability several years after the fact. Such is the situation which developed in this case.

This project concerned the construction of two bridges which carry the taxiways over the roadways at a major modern airport.

The design engineer for the project was retained by the owner to provide inspection of construction of the overall airport project. The design engineer then subcontracted the inspection of the taxiway bridges to another consultant.

The general contractor also let subcontracts—one for the fabrication of the steel girders which would be utilized in the construction of the taxiway bridges and a second to an engineer (other than the prime engineer and his subconsultant) to furnish inspection services during fabrication. Further, the prime contractor prohibited both the prime engineer and the prime engineer's subconsultant (who was to inspect the construction of the bridges) from visiting the fabricating plant and inspecting the girders during fabrication. His reasons for this act are not clear, but it was his contention that his agreement with the fabricator called for a purchase order for fabrication and did not constitute a subcontract. The prime contractor contended that the engineers were not permitted to intervene in a review of the work until the girders were delivered to the site. This arrangement proved to be unfortunate, because when the initial shipment of girders was delivered to the site and laid along the runways prior to installation, representatives of the subconsultant noted that the girders were so poorly constructed as to be patently unacceptable. Two problems were readily observable: (1) the improper alignment of the girders was such that upon laydown it was apparent that the tolerances for fit called for in the specifications had been exceeded, and (2) improper welding had been done on the girders.

In accordance with the powers conveyed by the subcontract, the subconsultant ordered the girders to be removed from the job site. Despite the patent deficiencies, the contractor requested permission to hoist several of the girders into place in order to convince the engineers that they were, in fact, properly manufactured. When this was done, it was even more apparent that the girders were entirely out of alignment and the entire shipment was rejected. To follow up on this rejection, and despite the above-noted prohibition against doing so, the engineers undertook a personal visit to the fabricating plant and observed similar problems with the remaining girders under fabrication. It was clear to the engineers that the plans and specifications were not being followed. Ultrasonic and x-ray testing of all fabricated girders was ordered. As a result, more than two-thirds of the girders which had been fabricated were deemed to have failed to comply with the specifications.

The contractor requested the engineers to permit repair of the defects by gouging out the defective welds. The subconsultant refused this request, contending that the final product would be of inferior quality. The prime contractor then offered a 5-year warranty instead of the original 1-year warranty for the girders. The subconsultant also rejected this offer as unacceptable, since the lifespan of the taxiway bridges was to be 30 years and the 5-year warranty did not cover the expected service life.

Subsequently, new girders were satisfactorily fabricated and the taxiway bridges were completed, but after a delay of more than a year. Liquidated damages were assessed by the owner against the contractor for a sum totaling $500,000.

The contractor thereupon commenced an action seeking recovery of the liquidated damages and various other claims relating to the design and construction supervision of the project. Uniquely, the suit was structured so that the contractor would be likely to recover regardless of how the evidence was developed. The theory of the contractor was that the delay damages assessed against him resulted either from improper fabrication by his subcontractor or failure of the firm he had hired to inspect fabrication and to do the inspection in a proper manner. The supplementary claims were to the effect that the assessment of liquidated damages by the owner was improper; the contractor developed the following areas of alleged wrongful acts by the design engineer and his subconsultant:

1. Arbitrary and capricious rejection of the steel girders
2. Improper rejection of the contractor's request to repair the bridges by rewelding
3. Failure of the prime and subcontract engineers to pass on to the owner the contractor's request for a time extension in order to complete the project
4. Faulty design of the proposed bridges, which caused unexpected and extensive costs in connection with pouring the concrete deck

The litigation produced an enormous flow of documentary evidence. After several months of negotiations, an agreement was reached whereby the design and inspecting engineers, for a nominal sum, were released from further participation in the litigation. The action continued further against the fabricator and the general contractor's consultant who had inspected the fabrication. Ultimately, a settlement was reached which resolved the contractor's claims and wherein the contractor recovered a substantial sum from the owner, the fabricator, and the consultants who inspected the fabrication.

TECHNICAL ANALYSIS

This structure consisted of all-welded, trapezoidal box girders topped by a concrete deck slab, roughly as shown in Figure 2-1a.

When this structure was received at the site, it was found that (1) the camber profile was not as specified and (2) the webs were buckled and the flanges were warped. As a result of the warping, the adjacent flanges did not meet properly, requiring filler metal or other treatment to form the required transition (see Figure 2-1b).

A second problem, found later, was that the web-to-flange weld, which had been specified to be a full-penetration weld (see Figure 2-2a), had actually been made using two fillet welds (Figure 2-2b) and that full penetration had not been achieved in a large percentage of these connections. The use of fillet welds was deemed unacceptable because of a lack of symmetry (bending stresses would result from nonuniform distribution of the transverse loads). Lack of full penetration was deemed to be unacceptable because of reduced fatigue resistance of the partial-penetration

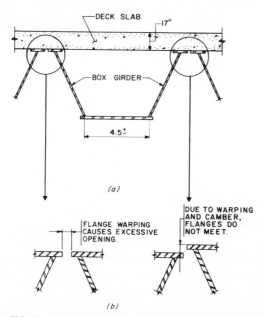

FIG. 2-1 Box girder topped by concrete deck slab.
(a) Design section. (b) Actual conditions.

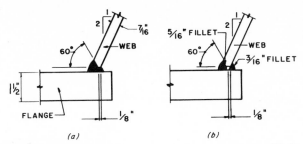

FIG. 2-2 Joint between web and flanges. (*a*) As specified. (*b*) As manufactured.

welds, even though they might have equal or greater strength under static loads.

Because of these and other, lesser defects, the girders were rejected. The ensuing controversy centered on whether, and if so how, they could be repaired satisfactorily.

The fabricator suggested the following repair procedures:

1. To correct the web-to-flange welds:

    a. Cut out and redo.

    b. In girders not yet fabricated, modify the connection detail, as shown in Figure 2-3.[1]

2. To correct camber:

    a. Block to remove tension stress.

    b. Heat bottom flange and lower portion of web in narrow strip to produce shrinkage and increase camber.

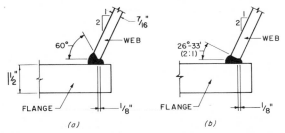

FIG. 2-3 Joint between web and flanges. (*a*) As specified. (*b*) Modification as suggested by fabricator.

[1] The anticipated advantages of this modification are (1) less filler material, hence less heat and distortion, and (2) more accurate cutting. Web plates are not absolutely flat, so that with a beveled cut the point of contact for a torch held at a fixed angle varies, resulting in a wavy cut and a variable volume of filler metal.

3. To correct web buckles:

   a. Jack to straighten.

   b. Small spot of heat to shrink out buckle.

The engineer's negative response to these proposals was based upon fears that potential distortions and locked-in stresses would ensue from the repairs. In particular, he was concerned with the following questions:

1. Would repair result in a substantial removal of parent metal and a consequent increase in the volume of weld metal to close the gap, as well as a resultant increase in heat and shrinkage?

2. Could a uniform gouge of the welds be made, or would it be nonuniform, resulting in stress concentrations and fatigue sensitivity due to variations in the volume of weld metal and heat input along the length of the connection?

In the end, the first set of girders was discarded and a second set was fabricated.

Comments

An interesting aspect of this case, as in Case 1, is the question of the effects of locked-in (residual) stresses on the ultimate strength of a member built of a ductile material and proportioned to develop yield without fracture. The reader is referred to the discussions in Case 1.

Another interesting aspect of the case is the concern with the fatigue strength of the flange-web connection. Checking fatigue life (allowable stress range) is important for structures subject to repetitive, varying loads. Fatigue life may be increased in the following ways:

1. Using a less sensitive type of connection (the allowable stress range for a full-penetration weld is about 50 percent greater than for a partial-penetration weld)

2. Reducing the stress range by using a stronger connection (reinforcing it)

Finally, this case illustrates the advisability of the designer specifying the nature and extent of the quality control which will be exercised during fabrication, particularly if ultrasonic, radiographic, or magnetic particle testing is to be required. First, fabrication is more costly, owing to delays for performance of these tests. Second, there is some argument that the custom in the industry is that if no such requirement is stated, welds will be acceptable so long as proper (prequalified?) procedures are followed and proper equipment used.

Another problem which developed in connection with this project was

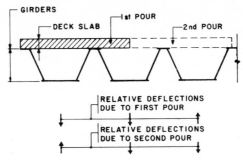

FIG. 2-4   Placement of deck slab.

that the type of design often dictates construction procedures and that it is desirable, where an unusual design concept is involved, for the designer to have in mind at least one well-considered method of construction; to indicate any known limitations to prospective contractors; and, perhaps, to indicate a suggested construction sequence on the plans. For example, Para. 4.4.1 of the American Concrete Institute (ACI) Standard 347-68 says:

> . . . special problems of forming with composite structures which should be anticipated in the design. Requirements for shoring or deflection control should be specified. With successive placements, deflection control becomes critical to prevent pre-loading reinforcing steel before imposition of the live load.

In this particular case, the contractor wished to place the concrete deck longitudinally, one lane at a time. The engineer insisted that the concrete be placed in transverse strips the full width of the bridge. The engineer's reason for this insistence is illustrated diagrammatically in Figure 2-4. The contractor's desire to pour longitudinal strips related to the paving machine he proposed to use. The anticipated effects of the longitudinal pouring sequence were (1) rotation of girders, (2) torsional stresses, and (3) web buckling.

LEGAL ANALYSIS

This was a case which was actively litigated by skilled counsel who were prepared to contest a high-exposure case with all the expertise at their command. As the case developed, it was learned that the contractor had prepared extensive materials in contemplation of litigation from the very outset of work on the project. This took the form of daily documentation by the contractor of any aspect of the project which could arguably result in delay damage claims, assertions of improper plans and specifications, including self-serving letters forwarded to all parties, and objections made during construction meetings. Unfortunately, unless design professionals are prepared to engage in a "paper war" on a daily basis, many unsubstantiated allegations of a contractor go unchallenged and hence are sometimes deemed to become the "law of the case."

Those familiar with the construction process will note that this is not an infrequent tactic engaged in by contractors. There is nothing inherently wrong with the contractor documenting the history of the project; nevertheless, it is up to all concerned to promptly respond to any factual distortions during construction so as to avoid such distortions from becoming accepted "fact" several years down the road.

This case also illustrates the need for early analysis of a situation and appropriate action to ensure careful review of any critical decision made during construction. The engineers in this case took extensive photographs before, and at the time of, the rejection of the girders. The hundreds of photographs vividly portrayed the poor workmanship in the construction of the taxiway girders.

It must be noted that an extraordinary amount of work was required in order for all counsel to develop the technical expertise needed to proceed properly with the discovery process. Countless hours were devoted to conferences designed to explain the different techniques associated with the design and construction of the taxiway bridges and their component parts. It was essential to a full understanding of the contractor's claims that each aspect of the design and construction project be mastered by the attorneys involved. In this way appropriate questioning and review of documentation in the possession of all parties could proceed with a critical eye toward the factual issues presented. Needless to say, any attorney who embarks upon the defense of an architect's or engineer's complicated claim without a similarly firm grasp of the technical issues involved is blindly entering the battlefield.

Finally, it should be noted that every step of the discovery process should enable the attorney to move on to the next logical step in developing a viable defense. Although this may sound simplistic, counsel can easily become so involved in the litigation process itself, that they fail to communicate with other parties to achieve a settlement. Had discussions with the owner and the contractor in this case not taken place, further litigation costs would have been entailed without improving the defense posture of the engineers. Moreover, an opportunity to resolve the litigation at an early juncture would have been overlooked, with the possibility that a similar opportunity might not have presented itself at a later date.

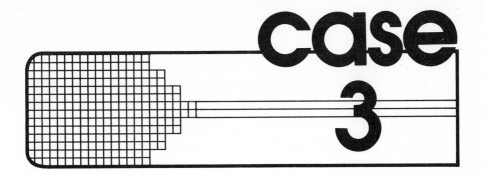

case 3

TYPE OF FACILITY   Building

TYPE OF PROBLEM   Cracking and distortion of masonry parapets

Significant Factors

A. Expansion of masonry units due to absorption of moisture
B. Improper detailing
C. Creep strain in supporting concrete framing
D. Poor workmanship

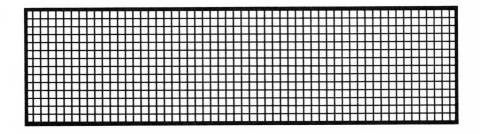

NARRATIVE

This project involved the design and construction of two dormitories for a college campus. The buildings were of twelve and three stories respectively. Construction of both involved a concrete frame (flat plate on concrete columns) with masonry and cavity (brick face and block backup) exterior walls. Each building was approximately 230 ft by 45 ft; the twelve-story building was topped by a 21- by 45-ft mechanical penthouse which divided the roof into two equal parts.

The initial problem which developed was one of distress in the masonry parapets of the twelve-story building (see Figure 3-1). Upon inspection, it was disclosed that where the parapet "panels" (sections between control joints) were isolated from the building corners and the penthouse (see Figure 3-2a), the parapets were leaning inward as much as 1 in. in the 3 ft from the outer joint where the membrane flashing occurs to the top of the coping. The parapet masonry was essentially crack-free, although other signs of incipient distress were apparent. These included signs of deterioration at the inboard horizontal mortar joint directly below the coping, which indicated a potential point of weakness for water entry. Some gaps also were noticed at a number of joints along abutting coping stones. A severe crack was also evident at the east end of the penthouse wall from the coping level to the middle of the wall at the first control joint.

At the lower building, the parapets also indicated inboard leaning to the same degree as that evidenced on the larger building. Extreme amounts of efflorescence were noticed on the walls below the flashed joint. Block

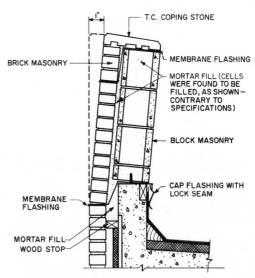

FIG. 3-1 Typical section of parapet.

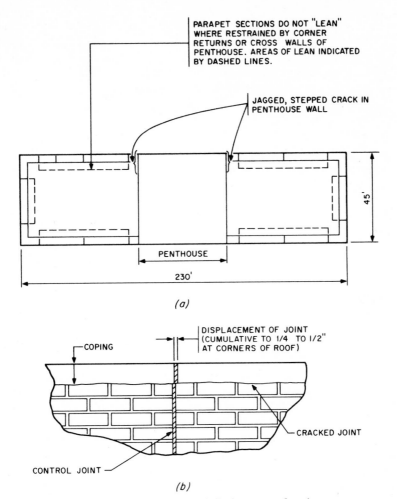

FIG. 3-2 (a) Plan of roof. (b) Gaps and displacement of coping.

cells were filled with water and it was noticed, upon removal of one of the coping stones, that the membrane flashing ended 1 to 2 in from the inside face of the parapet, thereby exposing block cells rather than covering the block completely and turning downward as a protective lip. Other problems included the fact that, instead of compressible filler running completely through certain control joints as detailed on the plans and specifications, mortar which had been used to set the coping stones made direct contact across the control joints.

The college called upon the architect, the structural engineer, and the contractor to conduct investigations into the cause of the cracking and to remedy the situation at their own cost. The contractor immediately retained an outside expert, who concluded that design deficiencies were the cause of the problems. Various bids for the remedial work ranged from $45,000 to $250,000.

The architect and the structural engineer, working jointly, retained an independent structural engineer to perform tests and prepare a written report detailing the causes of the failures. This report defined the major factors contributing to the masonry distress to be as follows:

1. The improper use of noncompressible filler under the relieving angles (see Figure 3-4)
2. The improper filling of the space from the toe of the relieving angle to the face of the brickwork with mortar (see Figure 3-4)

The report of the outside expert concluded that the cracks in the penthouse walls were directly related to these deficiencies. The walls were deemed unable to resist the inward forces generated by the attempted movement of the contiguous parapets, causing the joints to fail in shear and tension.

In regard to the three-story building, the distress was attributed to the narrowness of the installed membrane flashing below the coping. This deficiency, according to the experts, allowed water to enter the cells of the inner block wythe and penetrate the entire parapet construction. After several years of the freeze-thaw cycle, forces developed that caused displacement of the coping and parapet. While a similar construction deficiency (failure to cover the entire width of the parapet with the membrane flashing) was found throughout the taller building, damage from this deficiency was minimal, owing to the use of solid concrete (half) block in the top course directly below the coping. The expert's report concluded that had the walls and parapets been constructed in accordance with the contract documents, there would have been no masonry distress in either of the two buildings.

Meetings were thereafter held with representatives of the college and the contractor. The college representatives indicated that they would not take sides with either the design team or the contractor, preferring that the involved parties work out a solution whereby the damage would be corrected and the cost assumed by the responsible party or parties.

In this instance, no litigation ensued. Continuous meetings resulted in agreement by the contractor to assume full responsibility for repair of defects in the low-rise building. This left the problem of the parapet on the high-rise dormitory.

After numerous meetings, the contractor ultimately admitted that the problem with these parapets may have been attributable, in part, to improper construction.

The college had withheld $17,000 from the contractor under the retention provisions of the construction contract. Consequently, there was a financial stake on the part of the contractor to remedy the condition if he had any intention of recovering this retention. Finally, the general contractor agreed to complete the repair work on the parapet, and to submit a formal claim against the architect for the cost of this remedial construction. This claim sought the recovery of $63,000 for remedial work on the entire project, of which $43,000 related to the demolition and reconstruction of the parapets. To support this claim, the contractor pointed to an expert report he had commissioned which related the parapet distress to permanent moisture expansion of the brick utilized in the buildings. This was directly refuted by the expert retained on behalf of the design professionals. Accordingly, a decision was made to resist vigorously any attempt by the contractor to recover from the architect and engineer. A letter was thereafter written to the contractor on behalf of the architect, stating that his claim was denied in full and that he should expect vigorous opposition to any attempt to recover any of the costs incurred in the remedial work from the architect or structural engineer.

Apparently, this response, supported by expert analysis refuting the contentions of design error, was sufficient to preclude the contractor from pursuing any recovery via litigation. The matter ended with the contractor agreeing to assume all costs of repair.

TECHNICAL ANALYSIS

The principal concern in this case was the development of an inward "lean," or stepped displacement, of the masonry parapets (see Figure 3-1) amounting to as much as 1 in. in the 3 ft from the outer joint where the membrane flashing occurred to the top of the coping. The lean is shown in dashed lines and generally developed as an abrupt angle change in those parapet panels (sections between control joints) which were isolated from other restraints such as those at the corners of the roof or the penthouse walls (see Figure 3-2a).

A second concern was the development of a cleavage plane under the coping, a potential point of weakness for water entry (see Figure 3-2b). Longitudinal displacement of the coping (or parapet and coping) of ¼ to ½ in also was observed (see Figure 3-2b).

The contractor alleged that the observed movements were the result of expansion of the bricks due to the absorption of moisture. It is well known that ceramic bodies such as tile and brick (and concrete, as well) expand as moisture is absorbed and shrink as drying takes place. Displacement at corners of buildings and at intermediate changes in direction of walls and the associated cracking of masonry and of spandrel beams and columns are established occurrences (see Figure 3-3). Cases of up to 3 in of such expansion in a 200-ft length have been noted. This was cited

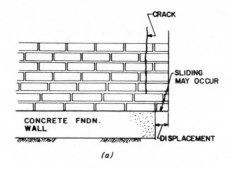

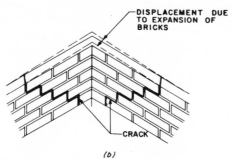

FIG. 3-3 Cracking due to expansion of masonry wall. (*a*) Typical displacement at corner of building. (*b*) Results of expansion at intersecting walls.

as the cause of the observed horizontal displacement of the parapets and of the coping. Added to this effect would be the creep strain in the columns of the concrete building frame, resulting in a relative upward displacement of the brick facing with respect to the concrete frame, in turn resulting in the observed tilt of the parapets.

The engineers' and architects' representatives agreed that the tilt of the parapets was caused by relative upward movement of the brick facing with respect to the concrete frame, principally due to creep strains in the concrete columns, perhaps abetted by temperature effects. However, they contested the idea of expansion due to moisture increase as the proximate cause, contending, rather, that poor workmanship and improper supervision had resulted in conditions which invalidated the function and purpose of the vertical control joints and the horizontal relieving joints which had been provided in the facing specifically to control the type of distress which had been observed. They showed, for example, the following:

1. The horizontal relieving joints had been built with an incompressible filler, rather than as designed (see Figure 3-4).

2. The functioning of the vertical control joints in the parapet had been invalidated by running the mortar bed under the coping stone through

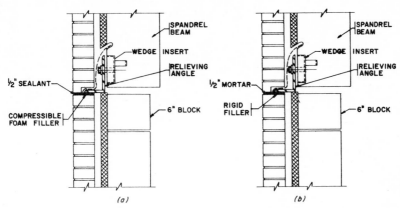

FIG. 3-4 Horizontal relieving joints. (*a*) As designed. (*b*) As built.

the control joint and by mortaring the control joints over the compressible filler. In some cases, "dummy" joints had been created by having mortar in the control joints, instead of a compressible filler, and applying the joint sealant over the mortar.

Comments

Designers must remember that what is obvious to them may not be understood by contractors and field personnel. Indeed, it is a frequent source of surprise to designers to discover how little builders understand of considerations that go into designs. In this case, the control and relieving joints (assuming the allegations of the architect-engineer, as described, are correct) were built in a manner which invalidated their performance. This has to be because the person who built them did not understand their function, and it raises the question of whether or not the contractor (or at least the supervisory personnel) should be expected to understand function. Is the designer looking for blind conformance to plans and specifications (in which case they had best be perfect), or is the contractor a partner in the work? Does an owner buy from a contractor only hours of labor and pounds of material, or is the owner entitled to expect a certain "know-how"?

The ideal answer to these questions is clear. The problem is how to get the contractor's input and understanding. One answer is prequalification of bidders. A more positive answer is the presence of the designers' representative on the job—to a sufficient extent to observe day-to-day operations. This introduces a major problem in the design-construction progression. Many designers are unprepared and/or unwilling to undertake field inspection of their work. Owners are unwilling to support the cost of such an inspection. Both attitudes are understandable, but unfortunate. The observant reader will detect a common thread running through many of the cases in this book: The contractor (neither maliciously nor with intent to cheat or defraud, but out of lack of understanding) doesn't do something, or changes something—and a problem results. In the end, all parties are losers. In the thrashings and flailings of self-protection, everyone is besmirched; recovery of damages, even where clearly justified, is seldom complete; and the costs of pressing or defending the suit fall upon all.

It is no answer to say that the contractor should simply follow the plans and specifications to the letter. First, the plans and specifications are seldom letter-perfect. More important, contractors (and the workers) have experience and good ideas, too, and should be encouraged to suggest time-saving or cost-reducing ideas, and not be precluded from doing so. But the only one fully qualified to judge the merits of these suggestions is the designer. It is a mistake to shut the designer out of the construction process.

Covert violation of the plans and specifications is another matter entirely. Even full-time inspection by an army of inspectors cannot hope to detect every design departure—if the builder is indeed intent on deviation or violation. What is needed is inspection, preferably by someone with at least a modicum of design background (or lacking this, at least in close liaison with the design office), and certainly by someone with the agility to get into close contact with the work and with enough interest and motivation to climb, slog through mud, and endure the occasional physical discomforts which accompany a field job.

A second lesson to be learned from this case is a repetition of the old adage that "the details make the job."

LEGAL ANALYSIS

Defending the architect in this matter involved negotiations with both the college, on the one hand, and the contractor, on the other. The college took the position that any responsibility for the remedial repair work belonged to the architect and the structural engineer and/or the general contractor. The college specifically refused to take sides or act as a mediator between the involved parties. From the standpoint of the architect, as agent of the owner, this posed a problem, since it gave impetus to the contractor's belief that the owner was not strongly standing behind the design team. Various entreaties to the college were unavailing, notwithstanding the fact that the architect had been commended on numerous occasions for excellence in designing the project.

It therefore became necessary to prepare a detailed analytical construct about the problems at hand and the factors which combined to cause the problems to the buildings. The immediate retention of an outside expert and coordination with the architect and structural engineer developed a dialogue which helped the design team re-create the causes of the failure.

Initial pressure had to be brought upon the contractor to admit that he had failed to use the proper sealant in the control joints and to agree that he would correct this deficiency at his own expense. The contractor was thereafter placed on the defensive, knowing that he had admitted liability to at least one failure. Subsequent meetings resulted in admission by the contractor that insufficient quantities of compressible filler (as called for in the plans) had been available and that he had resorted to asphaltic sealant with the intention of changing the sealant at a later point in the construction. It was pointed out to the contractor that a resident engineer for the architect had noticed, at a lower level of the building's construction, that the wrong sealant was being used. A written memo to the contractor called for the restoration of the correct sealant. It was ultimately discovered that this correction was never made by the contractor for the balance of the building.

This, of course, did create some problem for the architect, since, having once called the contractor's attention to this variance, there was no follow-up by the resident inspector to ensure further compliance with this request.

A decision also was made to have the outside expert for the architect detail his findings before the contractor, as well as meet with the expert who had been retained by the contractor. In this fashion, the carefully reasoned analysis and findings supporting the design were transmitted to the contractor and his expert for their consideration. This tactic also was designed to impress upon them the fact that any litigation would result in a vigorous defense on the part of the architect, in conjunction with an attack upon the obvious construction deficiencies, one of which had already been admitted.

During meetings with the contractor, it was pointed out that it was in his best interest to undertake the remedial repair work on his own, in order to avoid having the college put the repair job out to bid. This, undoubtedly, would have substantially increased the damages, a point the contractor chose not to disregard.

When the contractor did complete the repair work, and made an attempt to recover his out-of-pocket costs from the architect, a summary rejection of this claim by the architect was completely in order. Certainly, had the matter proceeded to litigation, a counterclaim by the architect and structural engineer for their own out-of-pocket costs and additional work to remedy the brick distress may well have been successful. Accordingly, the ultimate decision by the contractor not to pursue any loss he sustained in repairing the project appeared to have been the wisest course of action.

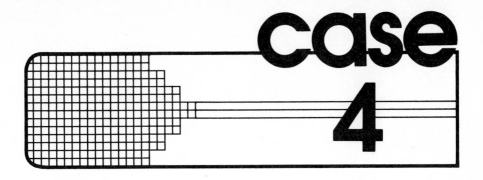

case 4

TYPE OF FACILITY   Bridge

TYPE OF PROBLEM
1. Structural failure of bearing seats
2. Excessive deflection

Significant Factors

A. Method for estimating shear strength of corbel (shear-friction vs. classic VQ/Ib)

B. Locked bearings and drag on sliding bearing seat

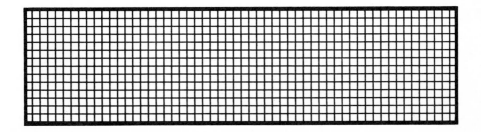

NARRATIVE

A simplified elevation of the bridge, illustrating features pertinent to the problems to be discussed, is shown in Figure 4-1. Aside from minor matters, two problems developed:

1. Cracking of corbel supports for precast girders
2. Sagging of cantilever girders

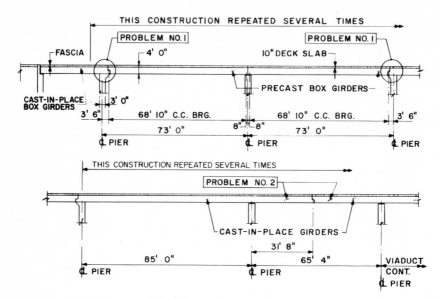

FIG. 4-1 Elevation of viaduct.

Cracking of Corbel Supports

Seven days after the erection of the initial set of precast girders, the concrete deck was placed. On that same day, the first crack in the corbel supports was noted. The crack, a hairline separation, was attributed to a condition of point loading caused by the fact that the box girders were warped and did not bear evenly on the bearing seat (elastomeric bearing pads had been installed between the girders and the seats). Work was halted. Structural calculations were made. The calculations verified the adequacy of the design, and a decision was made to continue the construction. However, as a precautionary measure, rock bolts were drilled up into the soffit of those corbels which already had been poured, in order to strengthen them (see Figure 4-2). Construction proceeded, but so did the cracking. As additional spans were set and the deck placed, the development of cracks in the corbels followed.

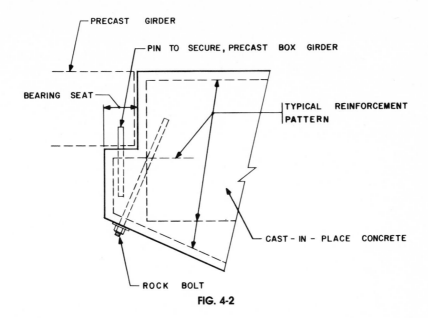

FIG. 4-2

The cracking still was attributed to the point loading, and the decision now was made to eliminate the rock bolts, to grind the bearing seats to match the soffits of the girders, and to install additional reinforcement in the corbels. This stopped the development of the cracks and the remaining construction was completed.

The initial hairline cracks continued to increase in size and to branch, however. In a 2-year period following erection, they eventually developed into a condition of obviously critical distress. Investigation revealed that moisture penetrating the roadway joints ran down onto the bearing seats, soaked into the cracks, and froze therein, implacably wedging the cracks open and causing the cracks to propagate. As a result, those spans in which cracking of the corbels had occurred were dismantled, the corbels rebuilt using more reinforcement, and new joint seals installed to prevent water penetration from the roadway. Those spans in which the corbels had been additionally reinforced before casting did not crack and did not require remedial work.

The matter of why the structural calculations failed to reveal the true nature of the stress condition and the failure of the rock bolts to improve the situation will be discussed under the section Technical Analysis.

Sagging of Cantilever Girders

The initial plans for this viaduct provided for a 16'0" cantilever overhang, whereas an overhang span of 31'8" is shown in Figure 4-1. The reason for the change was that, during initial stages of the construction, the owner, a municipality, had discovered that acquisition of the land to accommodate the pier at the cantilever overhang was a serious problem. The engineer thus was requested to redesign the structure in this area in order to alleviate the acquisition problem. This was done by increasing the cantilever overhang, but without an increase in depth of construction (because of restricted available depth between the roadway profile and required underclearance). Accordingly, additional reinforcement was provided to satisfy stress limitations.

It appears that, while stress conditions were satisfied, deflection calculations either were not made or their import was not recognized, because subsequent calculations indicated that theoretical deflection under dead load alone, and neglecting creep, would be over 4½ in. Large deflections (6 in), did occur, impairing the riding quality of the pavement. This condition remains in the structure, which is in active use.

TECHNICAL ANALYSIS

Failure of Corbels

An elevation (idealized) of the pertinent portions of the bridge is presented in Figure 4-1. The corbels were provided in the design in order to use uniform length sections of precast girders between piers which, for clearance, had to be at variable spacing. Details of the corbels (girder seats) are shown in Figure 4-3a. Failure developed as shown in Figure 4-3b. Col-

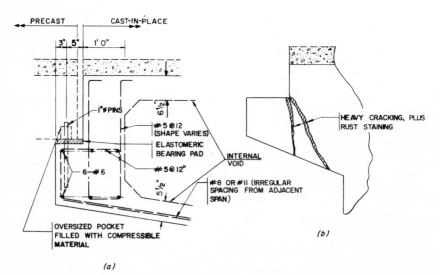

FIG. 4-3 Detail of corbel. (a) As-built detail. (b) Cracking which developed.

lapse did *not* occur, but substantial reconstruction was required to restore adequate safety, as described in the Narrative.

Various reasons were advanced as to the cause of this failure:

1. Inadequate reinforcement in the corbels. Inadequate thickness of the elastomeric bearing pads on which the girders sat contributed to a high longitudinal force which had to be resisted and for which the reinforcement which was provided was not adequate.

2. Anchor pins did not permit movement of girders on their seats and caused the edge of the corbels to be "pulled off" due to contractions resulting from temperature change, shrinkage, and creep.

3. Improper placement of reinforcement and anchor pins.

4. The girders did not bear evenly on the seats, causing local stress concentrations.

5. Poor concrete was used.

These arguments developed as follows:

Reinforcement of the Corbels

One might think that the design of the corbels, which have been used in structures for thousands of years, would be a well-established routine. It is not. Modern code provisions, e.g., ACI-318-71, prescribe two methods. The first method is based on the classic equation $f_r = VQ/Ib$. The other is the so-called shear-friction concept. The two methods can give widely different results (see Figure 4-4). The reason is shown in Figure 4-5, where

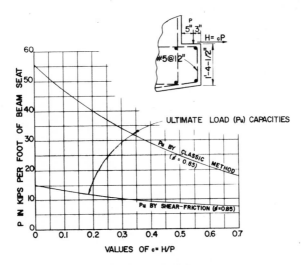

BEAM SEAT CAPACITY

FIG. 4-4 A comparison of two methods of designing corbels.

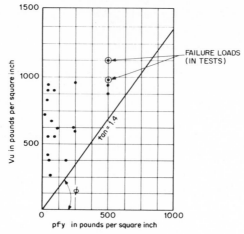

FIG. 4-5   Results of composite beam tests. (From *Journal of Structural Division,* ASCE, June 1968, p. 1493.)

the plotted points are stresses back-figured (by VQ/Ib) from failure loads determined in tests. The shear-friction method, using tan $\phi = 1.4$, represents a lower bound to the test data.

An essential of good detailing practice for corbels is to provide positive anchorage for the ends of the flexural and shear reinforcement and to distribute the shear reinforcement throughout the tensile zone of the corbel, as shown in Figure 4-6. Either of the details shown in Figure 4-6 would be an acceptable practice.

Also, the design of corbels must consider the longitudinal (drag) forces acting on the seat resulting from temperature change, creep, shrinkage, and, for posttensioned designs, elastic shortening.

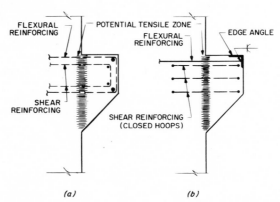

FIG. 4-6   Details for corbels. (*a*) Alternative 1. (*b*) Alternative 2.

The shear-friction hypothesis gives for this condition

$$V_u = (A_s f_y - H_u) \tan \phi$$

where
V_u = ultimate shear resistance
$A_s f_y$ = yield strength of reinforcement crossing potential crack
H_u = horizontal force at ultimate load
$\tan \phi = 1.4$, as a common value

The estimation of H_u requires careful consideration. In a bridge structure, bearings are provided to reduce these drag forces. Bearings in current use are *not* frictionless. Drag forces are developed. Because of corrosion, the buildup of dirt on bearing seats, the accumulation of ice derived from water penetrating the roadway joints, and the penetration of salt through these same joints, the drag forces increase with time and can reach unexpectedly high values. Locked bearings are a common occurrence. This is one aspect of bridge design where conservative design assumptions are warranted. For example, in this case, elastomeric bearing pads were used. The initial design thickness was ½ in. On account of tolerances in casting the girders and the beam seats, in some areas the actual pad which was installed consisted of 1 in of elastomeric material plus ½ in of asphaltic joint material for an overall thickness of deformable material of 1½ in. The girders which sat on the thick pads suffered markedly less, or no, distress. Where the girders sat on thin pads, cracking of the corbels was most pronounced. A common rule of thumb for the design of elastomeric pads is that the thickness of the pad should be at least twice the amount of the anticipated maximum movement required. In this case, a value of $H_u = 70,000$ lb calculated for the ½-in-thick pads would have been 13,000 lb if the pads had been 1¼ in thick (the calculated value of H_u to cause slip of the pads was approximately 18,000 lb).

The Anchor Pins

Expansion joints are a notorious source of trouble. They frequently lock, and the motion from two, or several, spans occurs in a single joint—the weakest one. The more elements in the joint, the more things that can go wrong. In this case, the anchor pins were a gratuitous embellishment. Unless there was so much movement that the bearing parts started to move off the seat (in which case the size of the pads and seats were incorrect), the pins were not needed, and if they were called upon to restrain movement of the 4-ft-deep girders, they were grossly undersized to do so. They would have been better omitted.

Furthermore, it appears that full consideration was not given to the tolerance to which the girders could be cast and to the changes in the length of the girders which would result from temperature change and shrinkage in evaluating the problem of threading the hole in the girder over the pin. As a result, in order to set the girders, some of the pins were set not *behind* the front bar of the reinforcing cage, but near the front edge of the corbel, where the slightest shear on the pin was bound to spall the edge of the corbel. Inaccurate placement of the reinforcement contributed to this problem.

Improper Placement of Reinforcement, Uneven Bearing of Girders on Bearing Seats, and Poor Concrete

These represent the universal problems and vagaries in the construction process and the need for inspection and vigilance in regard to critical items. In-depth inspection of all elements of a major construction job is often considered by owners to be an excessive expense. Hence, a "clerk of the works" (i.e., a single part-time, or full-time, inspector) is employed. In many cases, this is sufficient, but sometimes it is not. The authors have adopted in recent years the practice of stipulating on the plans those elements of the work which must receive in-depth inspection.

The jury's decision in this case indicated an opinion that these faults were not a primary cause of the observed failures.

Excessive Deflection of Cantilever Girders

Progressive deflection of concrete cantilevers is a common problem which, as the evidence in this case suggests, is sometimes overlooked. Stresses in the cantilever are readily determined; top reinforcement, only, is required, and it is only the top of the beam (i.e., the tension zone) which is flanged. With no flange or reinforcement on the compression face, creep deflection is maximized. ACI-318-71 contains the following expression for estimating creep deflection:

$$\Delta \text{ (total)} = \Delta \text{ (sustained loading)}^{(2-1.2\,A's/As)}$$

where
Δ = deflection
$A's$ = area of reinforcement on compression side
As = area of reinforcement on tension side

In a design involving a large ratio of live load to dead load, resulting in cracked section on the top surface, dirts gets into the cracks, preventing recovery of the live load deflection on removal of the live load, causing a continual increase in deflection due to "ratchet action." The freezing of water penetrating these same cracks produces a like effect.

Thus, conventional computations are likely to underestimate deflections of a cantilever in an exposed structure subject to heavy, varying live load. In the subject case, a dead load deflection of a little over $4\frac{1}{2}$ in was predicted by computation. A total, net deflection of about 6 in actually developed, impairing the riding quality of the roadway surface.

Comments

1. Basic economy of design takes place in the selection of the type of structure, the materials, and the principal dimensions. Normally, the details have little effect on the overall cost, but they are, by far, the primary source of difficulty. In this case, there is evidence to suggest that the designer, during the design stage, did not make any formal stress computations (at least any of record) regarding the corbels. He "eyeballed" the proportions and the reinforcement. Designers must work under the constraints of design budgets and deadlines. The need to eyeball some of the details is real and must be accepted, but it is well to have some tried and proven detail to adapt; otherwise, the risk is great. Of course, where a particular design is deemed to be innovative, all details should be reviewed.

2. Compare this case with Case 1. In both instances, it was the secondary stresses (stresses due to secondary forces—generally the forces due to restraint) which caused the problem. These forces cannot be ignored.

3. The sequence of events in the development of corrective measures is informative.

    a. Cracking of the beam seats was observed to have occurred as soon as the deck was placed, i.e., under dead load only. The designer, because of this cracking, reinforced some of the beam seats by installing rock bolts across the cracks (Figure 4-2) to supplement the reinforcement in the existing corbels. In some of the corbels which had not yet been poured, the shear reinforcement was increased (doubled).

    b. The bridge was completed and put into service. For 2½ years, the bridge was in use. Part of that time it carried truck traffic, although generally carrying a live load restriction of 6000 lb.

    c. In those corbels which had cracked despite the addition of rock bolts, the defects became progressively worse. The alleged cause was water—it penetrated the deck joints, entered the cracks in the beam seats, froze, expanded, and progressively forced the cracks open (ratchet action) until an untenable situation existed, requiring that the bridge be dismantled to rebuild the seats and put in new joint seals. It is well known that the expansion of water on freezing, if restrained, causes a substantial pressure. At 15°F this pressure amounts to about 15,000 lb/in², which is as much as the chamber pressure in some firearms.

d. In those corbels which had been additionally reinforced, no increase in deterioration occurred and the corbels remained in good condition. Once a defect developed, nature implacably sought out and attacked the point of weakness; where no defect existed, there was no weakness to attack. There appears to be a truism in structural engineering that if conditions are maintained above a certain level an extensive service life can be expected, but that if conditions fall below this standard, deterioration, progressing at a geometrically increasing rate, can be expected. The critical level is much a matter of opinion, but current criteria (ACI-318-71, for example) for crack control represent an approach to handling this problem. For exposed structures, these criteria and the development of unexpected defects may not be ignored. Defects must be taken seriously.

4. Design modifications and repairs which were used to correct and/or prevent cracking of the corbels included the following:

   a. Elimination of the anchor pins.

   b. Addition of rock bolts as per Figure 4-2.

   c. Change of deck to lightweight concrete.

   d. Increase in thickness of elastomeric pads.

   e. Increase in shear reinforcement.

   f. Sealing of deck joints to prevent water penetration.

   g. Pressure grout cracks—the girders were preloaded before grouting to open the cracks and facilitate grout penetration. Provision was made to squeeze the cracks closed after grouting by use of rods in drilled holes with takeup at the exposed end via plate and nut.

LEGAL ANALYSIS

This case can be clearly contrasted to Case 1, wherein there was early recognition and identification of a problem. In this case, unfortunately, the problem was not addressed until the structure had been completed, thereby increasing the damages drastically. While at this juncture it is conjecture, there is a probability that had the designer recognized that a problem existed which he could not fully explain and had he sought assistance, a remedy could have been accomplished prior to completion of the structure.

Another important lesson to be learned is that no project presents a simple task to the designer. In this instance, the designer's office was located in a state with a mild climate and he had extensive experience in the design of bridges of the type in question within his own state. This bridge, however, was being built in a locale which was subject to great temperature variations in short periods of time. The designer failed to take into consideration how the structure would react to these temperature changes. Ignorance of local conditions is not a defense against a claim of professional malfeasance.

Of passing interest, but of significant impact on the damages ultimately awarded, was the fact that this case involved a municipal funding with a limitation, as did other cases in this book. As the project had been completed prior to full recognition of the scope of the problem, litigation surely would ensue, since the only party who would be anxious to resolve the liability question would be the owner. The owner, obviously not enamored with the designer, had a remedial plan prepared which involved some upgrading of the structure. In order to fund this project, a local tax was imposed on motorists, which, when coupled with the inconvenience being experienced by local residents during the repair work, did not help in selecting an unbiased local jury.

Lessons to Be Learned

1. No design project is simple or without pitfalls.
2. When designing for locations in which the designer has no experience, greater care should be taken as to local practices, customs, climate, and regulations.
3. Early recognition and definition of a problem can result in a more favorable outcome of any dispute.

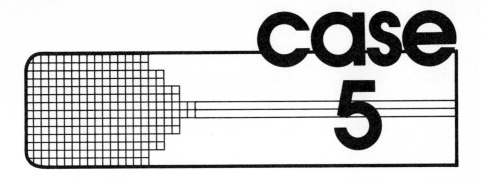

case 5

TYPE OF FACILITY   Septic system

TYPE OF PROBLEM   Backup (failure to function) of the system

Significant Factors

A. Improper use of a "preliminary" soils report

B. The design professional's posture in defense of a claim

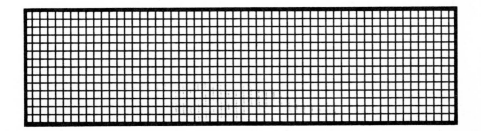

NARRATIVE

A firm of soils engineers was engaged to prepare a soils report for a proposed warehouse and office building on an undeveloped site. Their report, termed "preliminary," included the logs of five shallow borings and of seven test pits. The subsurface conditions were described in these logs broadly as "Topsoil overlying brown, fine to medium SAND, with some clay, gravel, boulders, and silt," or "brown silty coarse to fine SAND, trace of gravel." The report contained recommendations as to site preparation work (slopes), types of foundations for buildings (including bearing capacities), and some special problems such as ground water control during construction.

Following site preparation and landscaping, the building was constructed and occupancy commenced. After several months, the septic system used for sewage disposal from the office building backed up, flooding a portion of the building space. (A lesser problem, the sloughing of one of the cut slopes, also occurred but is not germane to this discussion.)

As a result of the backup and flooding, the owner commenced an action for damages against the engineer of record and the contractor. The engineer who designed the foundation and sewage system joined, as third-party defendant, the soils engineer, alleging reliance upon the preliminary soils report in preparing the design for the septic system.

Specifically, it was alleged that the soils classification and descriptions set forth in the soils report were inaccurate. The engineer claimed to have relied on these classifications and descriptions in deriving an assumed infiltration rate of 1.1 gal/ft^2 per day for discharge from the septic system. The soils engineer countered with the following defenses:

1. His report was intended to establish criteria for site preparation and for allowable soil bearing pressures, not for design of a septic system.[1]

2. The engineer designed the septic system without performing any percolation tests and therefore deviated from the accepted standard of care required in that locale.

3. The proposed buildings had not been sited at the time the report was prepared and the report was, accordingly, "preliminary" in nature.

4. The soils report has been used for a purpose (estimating percolation rate) for which it had not been intended.

---

[1] Review of that report indicates that the report was accurate in all that it said. It provided soil classifications—correctly; the level of ground water—correctly. It stated that the soil at the site contained good foundation material—which was correct; that the excavated material could be used for fill and that engineered fill using such material could sustain bearing pressures up to 2 tons/ft^2—also correct. It presented certain precautions and recommendations for handling ground water—all correct.

5. The engineer had not previously designed sewage or septic systems in the geographical area in question, and his misinterpretation of the soils data resulted, in part, from lack of familiarity with the soils of the area. The soil was a typical "till," with appreciable silt and a typically low permeability.

Upon investigation, after the action was brought against the soils engineer, it was learned that the classification "till" appeared in the text of the report, but not on the boring and test pit logs. This introduced questions of adequate presentation of facts by the engineer of record.

Significant (in regard to confirmation of the adequacy of the design) was the fact that a permit for the septic system had been issued by the local department of health. This raised the question as to whether or not the local health code required percolation tests (it did not).

Finally, a question was raised as to whether the design of the septic system had underestimated the flow which the system had to handle. This, in turn, prompted discussion as to whether the owner had provided accurate information on the number of persons who would occupy the new building. However, review of health and building department records revealed that the building had been using only 175 gal per day. Even if all of this was flow-through to the septic system, the system was being grossly underutilized.

A number of septic systems were in successful operation in the area.

During the litigation it was disclosed that the plaintiff, without advising the engineer, had another soils engineer do a detailed report which categorized the soils at the site as "moderately permeable" and suitable for a septic system.

The engineer for the health department who was interviewed also categorized the soil at the site, based on his experience, as "far in excess (more permeable) of the design criteria and therefore acceptable for a septic system."

What was going on? The facts clearly indicated that the system should work, but it wasn't working. After lengthy inquiry by the attorney for the soils engineer, the health department inspector recalled being told by construction workers that the buried sewer line leading from the distribution box to the septic tanks may have been crushed during the later stages of construction. Excavation revealed that this was, indeed, the case. The broken line was fixed in a matter of a few hours, and the system restored to good working condition. The problem clearly was that the line had been buried only 1 ft below the ground surface. Reconstruction of project records showed that the landscape contractor had been running a large bulldozer over the area after the septic system had been installed. It had taken some months for the sludge being carried out of the box to clog the crushed line; hence, the apparent delay before the problem arose.

TECHNICAL ANALYSIS

The technical aspects of this case are simple and need only simple commentary. The comments which come to mind relate to psychological, not engineering, problems.

A problem occurs. An accusation is made. The first reaction is one of self-defense. It is an unusual individual who undertakes to correct the problem first and worry about the blame later. Many readers will be aware of situations where more money was spent to fix blame, in preparing to litigate a problem, and in defending against such litigation than it would have cost to fix the problem in the first place. As a typical example, consider what happens when the concrete transit mix truck is late (a breakdown or a problem at the plant) and the partly completed pour is ready to take on an initial set. Are the workers busy forming keys and benches? Are they inserting scrap bars and steel to act as dowels to bond the old face to the new concrete when it arrives? Not usually! Instead, the inspector and the contractor are in animated discussion, there are angry calls to the batch plant, and the workers are standing around.

Another frequent first reaction is *mea culpa* (my fault). The construction process involves many steps and many parties. Many things can go wrong. Design codes provide substantial safety factors. An aggressive stance and the logical investigation of probable and proximate cause is more productive than self-condemnation.

LEGAL ANALYSIS

Significantly, the legal issues presented in a case of this kind center around the imperative need to conduct a thorough investigation of the project history so that the attorney can commence an early analysis of the problems on the project.

At the outset, it should be pointed out that the owner of the building did not sue the soils engineer, although there was a direct contractual arrangement between the two. The theory underlying the engineer's assertion that he relied upon the preliminary soils report as a basis for designing the septic system was contradicted by his testimony, which was taken at an early stage in discovery proceedings. The engineer clearly acknowledged that he had not previously designed a building in that portion of the state. His prior experience centered around another area of the state, where the soils classification fell on the other end of the geological spectrum.

Before a successful claim could be pursued against the soils engineer, a finding of liability against the engineer of record had to be predicted upon his erroneous design of the septic system. It became the conventional wisdom between the owner and the engineer to accept the premise that the septic system itself was improperly specified, and hence the need for the engineer to assert reliance upon the soils report classification.

Defense of the soils engineer was approached from a different standpoint. If, in fact, it could be shown that the system as designed was adequate for the geographical area and the needs of the building, then no liability for damage to the system or the flooding of the building could be sustained. This would, in turn, remove any possibility of a claim against the soils engineer. Retention of an expert at this stage can often remove any cloud of doubt in this regard.

In an effort to untangle the loose ends confronted by counsel for the soils engineer, a private meeting was arranged with the inspector for the health department who had reviewed the site planning and granted approval for the sewage system. After reviewing the file at the health department, it became quite clear that there was nothing out of the ordinary with respect to the design of the septic system. The question was put directly to the inspector: "In your opinion, is it customary to design a septic system for a warehouse and office building in this area?" The inspector responded directly and to the point: "It is a commonly utilized system and is perfectly acceptable as far as this department is concerned."

Curiously, none of the other parties had approached either the local building inspector or the health department for this information. It was the health department inspector who recalled conversations with construction workers leading to the information that the landscape contractor may have run a bulldozer over the line from the distribution box to the septic tanks. This information was then correlated with the nature of the damages sustained, and a soils expert confirmed that a blockage in the crushed line would be a logical reason for the backup.

At the same time, the attorney for the soils engineer secured, through a review of the documentation in the possession of the owner and contractor, a copy of a separate soils report which had been prepared at the request of the owner. This report, which had never been disclosed to the engineer or soil engineer, only came to light after examination of the files of the contractor. During the course of the litigation, the owner had not made any reference to the existence of such a soils report, which had been commissioned 3 months after the foundation soils report originally prepared by the soils engineer. The latter report was more comprehensive in nature and went well beyond the limited scope of the original soils report.

Clearly, this document provided all the necessary evidence to establish that the owner did not rely upon the original soils report as the basis for any portion of the design other than that relating to the foundation. The report stated that the soils located at the site were moderately permeable and hence would support a septic system. Interestingly, this second soils report was never furnished by the owner to the engineer during the design phase of the project.

Once this information had been developed, a meeting was held with counsel for the engineer to convince him that there was no basis upon which to predicate any liability against the soils engineer for a design error. In return for an agreement to dismiss the soils engineer, the materials produced from the files of the health department pertaining to the acceptability of a septic system were turned over to the engineer to assist him in defense of the claim by the owner. This information subsequently was of considerable value in enabling the owner and the engineer to resolve this matter through negotiation shortly before trial.

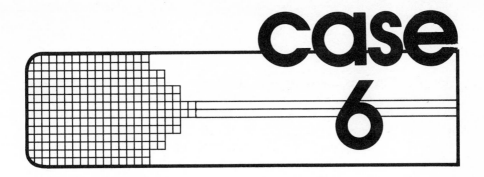

case 6

TYPE OF FACILITY   Hotel

TYPE OF PROBLEM   Inadequate reinforcement in concrete
slab

Significant Factors

A. Changes in plans

B. Rights in ownership and use of plans and
specifications

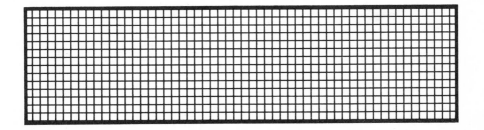

NARRATIVE

This case concerns the design for a high-rise hotel building which was adjoined by a one-story commercial area. The hotel owner had retained an architect to prepare the plans for the building. In turn, the architect had hired, under an oral agreement, a structural engineer who was to prepare the structural design drawings, but was not to participate in any on-site inspections, to check shop drawings, or to provide any other services after completion of the drawings.

A completed set of drawings was produced by the structural engineer. Subsequently, review by the owner and the architect resulted in a decision to attempt to reduce the construction cost by utilizing ultimate strength design instead of working stress design. The structural design was so amended by the structural engineer and the design drawings correspondingly changed. This did, indeed, result in a reduction of about 15 to 20 percent in the amount of reinforcing steel needed for the poured-in-place concrete floor slabs of the high-rise portion of the building.

The owner then retained a separate engineering firm to perform inspection services. Ostensibly, the reason was that the architect and the original structural engineer had their offices several hundred miles from the project site. These inspecting engineers were also responsible for reviewing the calculations and structural plans prepared by the structural engineer prior to construction.

During pouring of the second-floor slab in the hotel tower, the job was called to a halt by a representative of the local building department. The building inspector, during the course of the pouring, noticed what appeared to him to be an obvious lack of sufficient steel in the slab which was being poured. Investigation indicated that only four number 4 top bars were being placed in the column strip and that this represented less than one-third of the reinforcement (fourteen number 4 top bars) actually required. Modifications to the design were immediately made and incorporated in the construction of the remaining floors. Additional design changes were made to strengthen the second-floor slab.

A further check of the design drawings disclosed that the one-story section of the building also had several structural inadequacies. These were traced to changes made by the architect. He had relocated several columns. The revised column locations had been picked up on the structural drawings, but the size and reinforcement in the beams spanning between these columns had *not* been altered to conform to the needs of the altered (in some cases, increased) spans. Correction of these deficiencies also required redesign.

The owner, by now justifiably nervous, retained his own local engineer to inspect the entire matter on his behalf.

Needless to say, substantial delays in completion of the hotel and in securing a certificate of occupancy occurred, with a resulting loss of profit alleged in the litigation that ensued.

The action commenced by the owner named the architect, the original structural engineer, the local engineering firm responsible for on-site inspection of the reinforcing steel placement, and the contractor.

The investigatory stage of this litigation disclosed numerous overlappings of function and prevented a clear-cut formulation of liability against any one party. For example, it was determined at an early stage that the structural engineer, in developing the ultimate strength design, had derived a requirement for fourteen number 4 bars in the column strips in the slab. However, the set of drawings being used on the job showed a requirement for only four number 4 bars. Curiously, the lettering calling for the four number four bars was entirely different from that which comprised the entirety of the plans prepared by the original structural engineer. This raised the intriguing question as to who had made the change. The structural engineer contended that his plans had been altered by the architect—without the engineer's knowledge or permission. He cited the fact that the architect, during the period of performance of the work, had added a structural department to his (the architect's) firm and named an individual in that department who, he alleged, had changed the plans.

Handwriting experts were retained, and while they did determine that nobody in the office of the original structural engineer was responsible for the handwritten change on the drawings for the hotel, the changes to the drawings for the one-story structure had been made by the drafter employed by the structural engineer (which the drafter later admitted). It never was clearly determined who had been responsible for the error on the plans for the hotel.

As will be seen below, the resolution of cases of this nature often is based upon the experience of the counsel and experts retained to defend the matter. Errors were disclosed in the plans of the structural engineer which were unrelated to the reinforcing steel bar question. However, there was a disparity in the testimony of the various engineers as to the responsibility for picking up the obvious lack of reinforcement steel prior to construction. Moreover, the architect, in his deposition testimony, acknowledged liability in certain respects but indicated that he had insufficient assets and insufficient insurance available to pay the total amount of any judgment that might be taken against him.

The contractor took the position that he had complied with the plans and specifications furnished to him.

The result of the litigation was a settlement reached prior to trial. The owner agreed to a drastic reduction in his demand. This decision was based, in part, upon the fact that the architect was insolvent. In addition, the judge in the case, prior to trial, granted the resident inspecting engineer's motion for summary judgment. Thus, the original structural engineer remained as the sole design defendant at time of trial. In light of the acknowledged defects which existed in the design, and based upon an agreement among the parties to come to terms on the amount of damages sustained, such a settlement was in the best interest of all concerned.

TECHNICAL ANALYSIS

The problems inherent in a case of this nature are not of the technical variety. Rather, they relate to the practice of engineering in several important ways.

Plans are constantly undergoing revisions. Sometimes the changes continue even as the facility is being built. Coordination between the architect, engineer (structural, mechanical, electrical, et al.), and the builder is essential, but not always easy to accomplish. One important tool for preventing oversight and confusion is continuity of involvement in the work.

To have shop drawings checked by a firm other than the engineer of record is to assume a risk which needs to be justified by some special circumstances. The checking of shop drawings represents an opportunity for a "last review."

The desirability of keeping the same design team involved in the project throughout will be apparent. In the case under discussion, the design was changed just before going out to bid, from working stress to ultimate strength criteria. This change was made in order to realize the economy in reinforcement which ultimate strength criteria provide. Also, at or near this time, the architect had added an in-house structural department in order to expand the scope of his services. Although never established, the opportunity for confusion was obvious.

An interesting question is why the structural engineer was unable to produce a record print, a sepia, or a check print showing fourteen number 4 bars. His response was that the files on the job had been lost.

Finally, this case demonstrates that when a problem appears, it is well to take the opportunity to dig deeply to determine if other problems exist. Perhaps it is that minor problems result from some basic attitude of the designer or the supervisor, or from a lack of experience, or from an atmosphere of rush or confusion in the design office. Whatever the reason, experience indicates that problems tend to occur in bunches and on particular projects. Some common cause seems to be operative.

LEGAL ANALYSIS

There is much to be gained from analysis of the legal interrelationships between the parties to this project. Turning our attention first to the relationship between the architect and the structural engineer, it is important to acknowledge the difficulties that subsequently arose as a result of the oral agreement between the two. When a dispute arises between an architect and an engineer, whether it relates to additional services to be performed, to remedial design work requested, or to additional inspection obligations which are challenged as a part of the original understanding, without a written document to verify their understanding, misapprehension between the parties is more likely to prevail. Certainly, in reconstructing the history of any project after a claim arises, having the guidance afforded by a written contract enables counsel to understand the intention of the parties insofar as their relationship and function on the project is concerned.

The transmittal of documents between design professionals also is the subject of concern. Without doubt, transmittal letters covering such documents and specifying what is included within a given package should be maintained in a separate file by both parties. Carefully retained originals and copies of all working drawings, calculations, and other notes should be maintained for a period of several years depending upon the appropriate statute of limitations in each state. Reproducing this information at a later date during the construction process or after completion of a structure can substantially assist counsel and the design professional in re-creating the facts and circumstances surrounding a particular time period during the course of the project.

It is, of course, highly unusual when the plans of a building are changed without the knowledge of the principal party involved. In this situation, investigation did disclose that a draftsman in the office of the architect may have worked after hours attempting to do his own calculations to confirm the ultimate strength design of the structural engineer. Apparently, in the process of this "classroom exercise," the draftsman may have forgotten to restore the actual figures for the structural reinforcing bars to the plans after he had finished his own calculations.

Another interesting aspect of this case involved the settlement negotiations. Design professionals often inquire of their attorneys in regard to how the percentage contributions of each party are determined in reaching a negotiated settlement package. Quite often there is a clear-cut disparity in terms of the liability of each party. Other times there is such a complete overlap of function as to warrant an equal apportionment among the defendants.

On occasion, an intervening factor occurs prior to settlement, which may result in a total distortion of the settlement arrangement. Such an occurrence happened in this case when, during the final months of negotiating a settlement immediately prior to trial, the architect advised that he was without sufficient assets to pay any substantial judgment. In fact, the architect's attorney produced a certified audit showing that were he to pay more than a certain amount of money, he was prepared to go into bankruptcy and thereby forgo being a party to the settlement at all. As a result, although the architect would have been a prime contributing source to this settlement, he became a secondary source of funds. In order to reach the requisite amount demanded by the owners of the hotel, the structural engineer was required to advance a sum in excess of that which he may reasonably have been expected to contribute. The reason for making this additional contribution was to avoid a trial which would have kept the structural engineer as the sole, and hence "target," defendant for damages far in excess of that which could have settled the case prior to trial.

It is this type of exigency which requires both counsel and design professional clients to remain flexible during settlement negotiations and avoid a fixed position which precludes unforeseen possibilities.

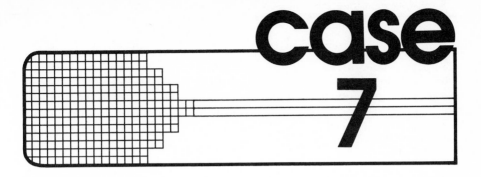

case 7

TYPE OF FACILITY   Parking garage

TYPE OF PROBLEM   Detailing of reinforcing bars

Significant Factors

A.   Lack of adequate detail in the plans

B.   Lack of adequate shop-drawing review

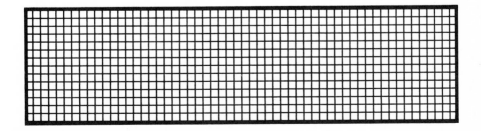

NARRATIVE

The building in this case is a cast-in-place, flat slab, concrete underground parking garage. The construction included two underground parking levels—a base slab and an intermediate deck—with a roof, i.e., three levels in all. The roof was designed to support the weight of a 5-ft cover of soil with associated landscaping and plantings, as well as pedestrian and service traffic.

At approximately 5:30 P.M. on a Tuesday, a 120- by 120-ft section of the roof collapsed. The roof fell onto the intermediate level, and then to the bottom level, carrying the intermediate level with it. The portion of the structure which collapsed had been poured and in place for approximately 60 to 90 days prior to collapse, and the contractor had been engaged in placing earth fill on the roof on the day that the collapse occurred. Approximately 2 to 3 ft of fill had been placed.

Fortunately, the collapse occurred after the workers had gone home, so that no one was injured. In fact, no one saw the actual failure.

The nature of the collapse can be seen by reference to Figure 7-1, which has been simplified for purposes of illustration. The collapse occur-

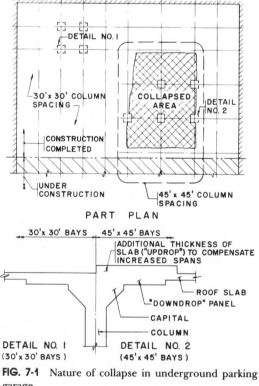

FIG. 7-1 Nature of collapse in underground parking garage.

red in the area of the structure where the spans had been increased from the normal 30- by 30-ft module to a 45- by 45-ft module to accommodate special use of the area. Here, the slab thickness (and strength) had been increased by the addition of "up-drop" panels (turned up to maintain headroom in the occupied space below; see Figure 7-1, Detail no. 2). As can be seen from Figure 7-1, the collapse occurred around the perimeter of the up-drop panels, which remained in place amid the collapse.

Investigation revealed that the cause of the failure was a deficiency in detailing the reinforcing bars in the up-drop panels (see Figure 7-2a), and much attention was devoted to establishing responsibility for this deficiency. In accordance with normal practice, the bar detailing had not been done by the design engineers, but by a firm specializing in the detailing of reinforcing bars. The detail in question was not shown on the plans but was developed by the detailer, allegedly in consulation with the designer. The detail (shop) drawings were checked by the design firm without detecting the problem and were reviewed by the owner (a public agency), also without detecting the problem.

The design engineer was not engaged for inspection of construction,

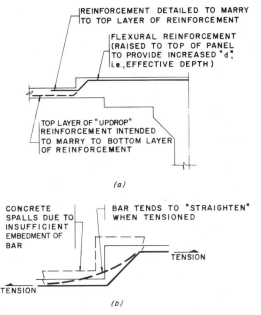

(a)

(b)

FIG. 7-2 (*a*) Error in detailing reinforcing bars in undrop panels. (*b*) Consequences of error in detailing. Step 1, bar tends to "straighten" when tensioned; step 2, concrete spalls due to insufficient embedment of bar; step 3, bar does "straighten" when relieved of confinement by concrete; step 4, "straightening" in effect relieves tension in bar, which causes cracking and collapse.

but did make some visits to the site. There was some debate over whether he should have noted the problem during those visits, and some evidence was introduced that it had been pointed out to him by field personnel.

Allegations were also made that poor quality concrete had been used in some areas of the structure, causing excessive in-plane stresses (shrinkage and temperature). These allegations were viewed and rejected as proximate causal factors.

Corrective work consisted of rebuilding the collapsed areas and the remaining areas containing the improperly detailed bars. Rebuilding was to the same design, but using higher strength concrete and correcting the improper bar detail.

TECHNICAL ANALYSIS

The technical point illustrated by this case is presented in Figure 7-2b. It seems a small thing to have caused several millions in damages, but is dramatic proof, as stated in Case 2, that the "details make the job." The following observations are offered:

1. The task of checking shop drawings at times gets assigned to the least experienced junior staff. Often, it is treated as a subprofessional task and/or a job useful for training or for "fill-in" assignment. Actually, it is the engineer's last chance to review his or her work.

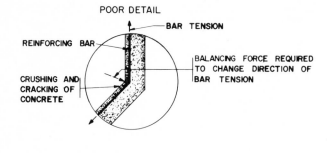

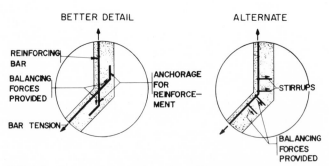

FIG. 7-3 Cause and prevention of crushing at a reentrant corner.

2. Detailing of reinforcing bars is a task usually left to the contractor, as part of the bid for the work, and who accordingly retains the least costly service. Many an owner, who would not think of buying design services on the basis of price alone, accepts this practice—possibly unaware of the risk involved.

3. This case is also a good illustration of the "zipper effect." As failure occurred in one area, the load was transferred to other, stronger areas, which in turn became overloaded, causing progressive collapse.

4. Finally, Figure 7-3 is taken from a book written in 1965 (Sidney M. Johnson, *Deterioration Maintenance, and Repair of Structures*, McGraw-Hill, 1965). It is curious how the same mistakes get repeated.

LEGAL ANALYSIS

In its simplest terms, this case turned on the respective duties of a designer and a fabricating detailer. The supplier of the steel argued that the bar in question should have been detailed on the plans and specifications and that the omission of such a detail constituted an error on the part of the designer. The designer contended that the omission of the detail from the plans was not causative in the collapse, but rather the causative factor was the detailer's assumption of the role of designer in assuming the location of the bar.

Applicable standards were researched and support was developed for the proposition that, under the circumstances presented to the detailer, he was bound to notify the designer of the omission and not to assume the location. Of course, the designer's position was weakened by the fact that in checking the shop drawings he did not determine that the detailer had improperly located the bar. This focused attention on the still unresolved, but greatly debated, legal issue as to the responsibility of the designer in checking the shop drawings. No useful purpose would be served by attempting to digest in this work all the legal opinions which have been written on that issue. Suffice it to say that the law is in a state of flux and that the issue is not one which can be commented upon with brevity or certainty. What is important to the designer is that the legend on his shop drawings stamp reflect the most recent pronouncement on the issue,* and to this end, he should consult with his attorney on a regular basis. Although not germane to the issues in this case, the same comments

---

* See, for example, Article 1–5–13 of AIA Document B141, Owner-Architect Agreement, July 1977, which provides as follows: "The Architect shall review and approve or take other appropriate action upon the Contractor's submittals such as Shop Drawings, Product Data and Samples, but only for conformance with the design concept of the Work and with the information given in the Contract Documents. Such action shall be taken with reasonable promptness so as to cause no delay. The Architect's approval of a specific item shall not indicate approval of an assembly of which the item is a component."

hold true for the language contained in the designer's standard form contracts.

In this case, as in Case 1, the project budget was fixed by legislative pronouncement. It was therefore necessary to make an early determination as to responsibility and to provide funding for the remedial work. Cooperation did exist between the parties, so that the contractor agreed to continue working, with a revised set of plans, on the understanding that a reasonable resolution would be achieved prior to the existing funds being exhausted. The owner, likewise anxious to minimize its own delays, was open to such a proposal and the matter was eventually resolved.

Resolution of the matter was complicated by the introduction of an extraneous issue. As has been noted in the Narrative section, the designer was not engaged in inspection of construction but did make occasional visits to the site. During one of the visits, the site engineer was asked to view the steel placement at the location of the deficiency. The engineer, not really conversant with field conditions, faced with a mass of steel bars, muttered some response to the supervisor which seemed appropriate at the time. Unfortunately, no record of this site visit was made by the engineer and it was ultimately claimed by the field personnel that approval of the designer as to the placement of the reinforcing bar had been obtained during this site visit. As is often the case, gratuitous advice or the providing of services not required by contract can, in confrontation situations, either confuse the matter or mitigate against the designer's position.

Lessons to Be Learned

1. Shop-drawing review is an important aspect of the construction process and should be treated as such. Logs should be maintained of the dates shop drawings are received, the dates they are checked, the action taken, and the dates they are returned to the contractor.

2. The designer's shop-drawing approval stamp should contain a proper description of the extent of the review with appropriate disclaimer language. The shop-drawing review stamp language should be reviewed with counsel on a regular basis to insure that it is conformance with the latest legal decisions on the designer's duty in reviewing shop drawings.

3. Gratuitous opinions and advice should be avoided. A designer should not express an opinion unless certain of the answer and prepared to defend it in the event of litigation.

4. Records should be kept of all visits to the site with proper commentary in regard to dialogue and transactions.

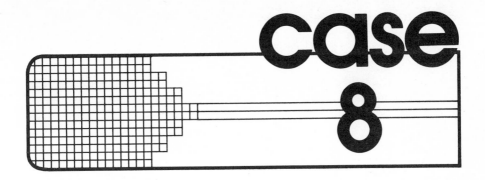

case 8

TYPE OF FACILITY   Parking garage

TYPE OF PROBLEM   Inadequate shear strength in concrete
     beams

Significant Factors

A. Diagonal tension failure

B. Increase in shear stress due to torsion

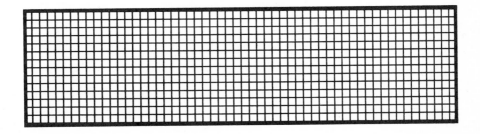

NARRATIVE

The structure in this case is a six-level, above-grade parking garage, 180 ft by 300 ft in plan, built using precast concrete units (not prestressed), roughly as shown in Figure 8-1.

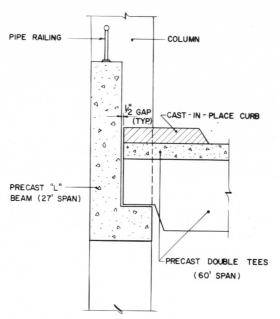

PIPE RAILING

COLUMN

½" GAP (TYP.)

CAST - IN - PLACE CURB

PRECAST "L" BEAM (27' SPAN)

PRECAST DOUBLE TEES (60' SPAN)

FIG. 8-1 Diagram of precast concrete units used in building the parking garage.

About 2 months after completion, cracks were noted in the precast L beams, of sufficient severity to cause alarm. Investigation revealed a deficiency in shear strength due to the reduced section associated with a dap cast into the beams to accommodate connection to the columns.

Corrective action consisted of adding external shear reinforcement in the affected areas of the affected beams (see Figure 8-2).

This claim was settled for about $70,000.

TECHNICAL ANALYSIS

The observed cracking (in the L beams) followed the 45° pattern classically associated with distress in diagonal tension and, indeed, was found to be the result of inadequate shear (diagonal tension) strength resulting from

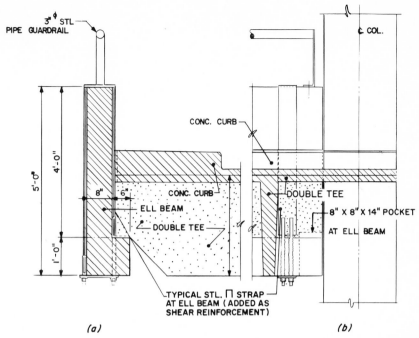

FIG. 8-2 External shear reinforcement added to the precast concrete units. (*a*) Typical section. (*b*) Typical elevation.

the decreased depth at the daps cast into the units. The following points shall be noted:

1. The shear stress due to end reaction was increased by the torsional rotation caused by the eccentric bearing of the double tees on the L ledge. It was the torsion that ran the shear stress "over the limit." It appears that this effect had been overlooked in the design. Indeed, no calculations seem to have been made of the web stresses at the dapped section of these beams.

2. The designer may have relied on the precast detailer to develop the end details for the beams. The detailer did not detect the problem.

Repair consisted of adding external shear reinforcement as shown in Figure 8-2. Epoxy was injected into the cracks to rebond the concrete surfaces.

Once again, it is emphasized that the details make, or break, the job. This can be a problem. The correspondence in this case suggests numerous last-minute changes by the owner. One can readily see that details might be given cavalier treatment in a "charette" situation. Those who engage engineering services should understand that they hazard quality and economy when they require, or cause, the work to be done without adequate time for its careful execution.

LEGAL ANALYSIS

This matter was settled short of litigation, thereby resulting in a substantial savings in legal expenses. The matter was settled on reasonable terms, owing in part to proper early attitudes when the problem was first discovered. Early recognition of a problem is most often the key to a successful resolution of the dispute short of litigation. Having discovered the problem and analyzed it, the owner can be approached with a reasonable explanation and a suggested remedy. Ignoring the problem or denying responsibility when responsibility exists creates an immediate adversarial relationship between the design professional and the owner. Once that occurs, their respective positions become more adverse by the mere operation of human nature. The owner will feel betrayed by the designer and seek counsel from another designer. The second designer, recognizing that there is a conflict, may be overly cautious, not wishing to be charged later with negligence when litigation ensues. Thus, the remedial scheme may be more conservative and have a higher cost attached. The damage to the owner is also increased by virtue of his having to pay a second consultant. The antagonism between the owner and the original designer is then increased.

By accepting responsibility for an error and discussing it forthrightly and reasonably with the owner, the design professional will usually keep the owner's confidence. So long as that confidence is maintained, the owner will most likely accept a remedial scheme devised by the designer. By preserving cordiality with the owner, the matter is ready for reasonable negotiations toward settlement. Avoiding the courtroom saves the designer both time and money.

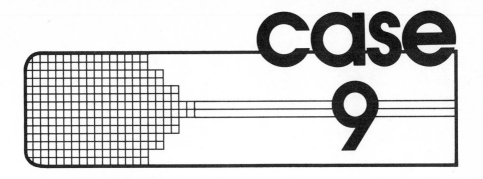

case 9

TYPE OF FACILITY Prefabricated metal building

TYPE OF PROBLEM Collapse during rain storm

Significant Factors

A. Ponding due to:
1. Excessive flexibility of framing
2. Clogging of roof drains
3. Incorrect location of roof drains

B. Structural action assumed in design differed from reality because of neglect of stiffening effects of masonry walls

C. Design for uniform live load, neglecting effects of nonuniform loading conditions

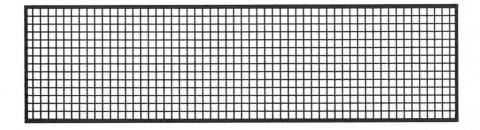

NARRATIVE

In 1964, a manufacturer and designer of prefabricated (steel frame) buildings sold such a building to its local dealer (a construction company) who then erected the building for the customer as a plant facility.

After 10 years of service, and following a rainfall totaling only 1½ in. in a 6- to 8-hour period, a portion of the roof of the building collapsed. The collapse was progressive, one frame following another, involving a total of twelve frames in a period of approximately 10 minutes. The plant was closed for 3 days and was only partially operational the following week. Damage was also sustained by the contents of the building.

An investigation performed by engineers retained by the owner concluded that the building had been structurally inadequate to sustain the roof's design live load of 30 lb/ft² and, more importantly to the development of the collapse, was unable to sustain unbalanced loadings of even lesser magnitude. It was demonstrated that such unbalanced loadings had occurred because of ponding of rain water on the roof.

Investigation by engineers retained by the manufacturer of the building concluded that the design *was* adequate to sustain the uniform design load of 30 lb/ft², but agreed that it was unable to support unbalanced loads of lesser magnitude.

Suit was brought by the owner against the manufacturer and the erector of the building. Total damages claimed were $200,000.

In defending against this claim, the attorneys for the manufacturer first alleged that at the time of purchase and erection of the building the state of the engineering art (standards of the Metal Buildings Manufacturers Association) generally did not take into consideration ponding and that understanding of this matter was not publicly reviewed in the engineering trade journals until 1965. Other questions concerned the following:

1. Identifying who had established the design loading criteria for the building
2. The influence of subsequent tenant changes on the adequacy of the structure
3. The adequacy of maintenance (cleaning) of roof drains, a failing which allegedly contributed to the ponding condition

Further investigation also revealed that in 1966, i.e., about 2 years after erection of the building, another, similar, building by the same manufacturer, which was erected for the same owner in a different state, had collapsed under somewhat similar conditions. This collapse also had been traced to a condition of ponding and the resulting unbalanced loading. As evidence that the manufacturer was aware of this deficiency, it was pointed out that the manufacturer had, at its own cost, installed additional purlins in the roof of that building and at two similar buildings in order to stiffen the roof, reduce ponding, and hence the potential magnitude of unbalanced loading. It was arguable that the manufacturer had been put on notice of a deficiency in its design. Further argument was made by engineers for both the owner and the manufacturer that the additional purlins were not a proper solution to the problem of unbalanced loads.

Settlement was effected for $150,000, partly in cash and partly in remedial work. Both the cash settlement and the remedial work were provided by both the manufacturer and the dealer-erector.

As an interesting aside to the issues in this case, it appears that the plans for the building bore the seal and signature of a local, retired engineer who, for a nominal sum, had affixed his seal and signature without having reviewed the plans. Theoretically, this engineer was liable for the deficiencies of the design. Being uninsured and possessing no independent wealth to pursue, the parties decided not to involve him in the matter.

TECHNICAL ANALYSIS

Two different conclusions as to the proximate cause of the failure were advanced. The differences between them never were resolved and both theories are presented. Both investigators agreed, however, that the basic cause of collapse was unbalanced roof loading, caused by ponding of rainwater, and that the structure was capable of supporting loads of even greater magnitude than those to which it had been subjected at the time of collapse—had the loads been uniform.

First Investigator

A schematic diagram of the structure is presented in Figures 9-1 to 9-3. Only half the roof collapsed. The collapsed portion was that indicated in Figure 9-2.

Analysis was made for live load only in the two interior bays, next to the firewall. It was concluded that failure resulted from web crippling (due to flexure) of columns along Column Line 3, as shown in Figure 9-4. One of the main factors was the presence of the masonry wall, which was built tight to the underside of the steel and, in effect, increased the fixity of the joint, by preventing rotation of one side of the joint. The deformations due to the web crippling in turn induced moments and rotations at the beam end-plate splices which snapped the bolts and led to collapse.

A second important factor was the ponding of water on the roof and the inherent flexibility of the structure. The roof sagged away from the high points defined by the supporting masonry walls. The roof drains were located near the masonry walls (the theoretical low point, defined by the

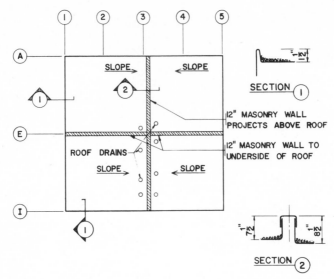

FIG. 9-1 Schematic of roof plan of the prefabricated metal building.

roof slope). However, because of deflection of the roof framing, this was not the true low point. The low point was out in spans 2-3 and 3-4 and, because of the aforementioned restraint to deflection offered by the fact that the firewall supported the beams in span 3-4, the low point actually occurred in span 2-3.

Failure to maintain the drains (clogged drains) and the existence of the 8-in projection of the firewall above the roof, in effect forming two separate water compartments, were also factors. Once ponding started and

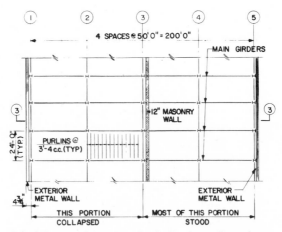

FIG. 9-2 Partial framing plan indicating collapsed portion.

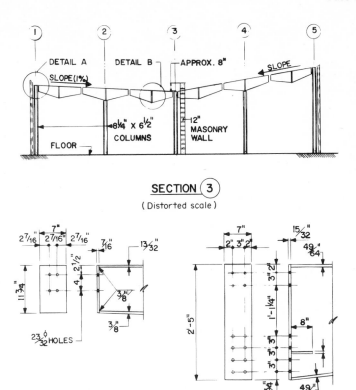

SECTION ③

(Distorted scale)

DETAIL Ⓐ   DETAIL Ⓑ

FIG. 9-3

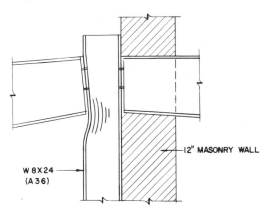

FIG. 9-4 Web crippling of column along Column Line 3.

the roof began to deflect, the drains were increasingly located away from the drainage low point. The effect was cumulative. The report's conclusions stated that the collapse would not have occurred had *all* the roof drains been working and had the masonry walls not been present. It was noted that the roof had stood in an even heavier rain storm which had occurred some years earlier.

Second Investigator

The second investigator also concluded that the roof did not entirely drain to the intended low point where the drains were located and agreed that ponding was the proximate cause of the failure. He also agreed that the exact depth of ponding could not be determined. However, his investigation and, reputedly, eyewitness reports to the collapse indicated that the columns with the crippled webs, which were found in the collapse debris, did not come from Line 3, but rather from Line 1. The failure mechanism shown in Figure 9-5 was postulated. Based on moment-deformation (M-ϕ) tests of the end-plate splices, calculations indicated that moments developed under the nonuniform loading condition postulated in Figure 9-6 caused the joints at B, D, and E to act as hinges, i.e., to have no effective resistance to rotation. Bolts in these joints failed, section DE fell, and the release of this load then caused section AB to fall.

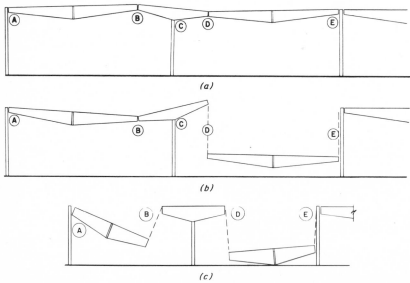

FIG. 9-5 Postulated failure mechanism causing collapse. (*a*) First stage of collapse. (*b*) Second stage of collapse. (*c*) Third stage of collapse.

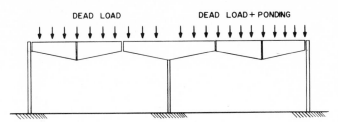

FIG. 9-6 Postulated nonuniform loading condition which caused collapse.

Discussion

The above dispute between experts is presented without comment. The remedial measures which were taken in the sections of framing which had not collapsed speak for themselves:

1. Main girder splices at *B* and *D* were reinforced by adding top and bottom cover plates to increase moment capacity.
2. Diagonal braces were added to stay the bottom flanges of the main girders against lateral buckling failure.

LEGAL ANALYSIS

As will be noticed from the above technical discussion, the manufacturer of this prefabricated building also could be construed to have been the engineer of record. However, for purposes of defending the manufacturer, a strong argument was made that it had acted solely as a supplier of materials, furnishing parts in accordance with minimum requirements set by the dealer. There was general agreement by the expert engineers that the building, as furnished, was adequate to support the balanced load specified by the dealer.

However, a similar building manufactured and installed for the same owner in another state did collapse in 1965, was repaired by the manufacturer and the dealer, and again collapsed in 1966. Remedial work undertaken by the manufacturer and the dealer consisted of purlin additions to correct an apparent ponding condition. This prompted the manufacturer's engineers to review other similar buildings constructed for the owner in different states, including the building which is the subject of this case. In all such instances, additional purlins were added to the roof. According to the representatives of the manufacturer, the additional purlins were added solely to correct the continued deflection of secondary members (ponding) and not to cure problems associated with clogged drains or other causes for the collection of water on the roof.

Arguably, the manufacturer could have been deemed to be aware of a design defect concerning the projection of the firewall above the roof as an aggravating factor. However, this was disputed by the manufacturer's representatives. The essential question in the defense of the manufacturer was, therefore, whether its engineers, who personally inspected the collapsed plants in 1965 and 1966, knew or should have known of the ponding effect on the "primary" members, as well as on the secondary members. Further, a question was raised as to whether there was a duty for these engineers to physically examine the building to determine whether remedial measures should have been taken at that time, as well as on other buildings similarly designed.

Finally, a problem was associated with the fact that the manufacturer did make necessary corrections in other plants which had exhibited similar ponding conditions without making similar changes to the plant in question. The owner argued that the corrections made on the other plants located around the country would have averted a collapse at its plant. In effect, a legal issue was raised concerning whether the manufacturer assumed a duty to make all necessary corrections to each of the plants similarly designed, and, if that duty existed, whether it was breached by the failure to properly perform all necessary remedial repair work on buildings similarly exposed to such hazards.

A significant obstacle to obtaining a settlement of this case was the desire of the manufacturer to secure a release not only for future damages to the plant in question, but for all other similar buildings supplied to the owner. The owner, needless to say, was reluctant to give such a release, citing its belief that the other buildings were structurally deficient and required a substantial expenditure in order to be made safe.

During the course of settlement negotiations, it was agreed by the parties that any release would be limited to the building in question, and while no release would be given by the owner with respect to any alleged deficiencies, no further payments would be made by the manufacturer or the dealer. For purposes of settlement, the manufacturer agreed to perform all necessary engineering services and to provide materials to ensure that the structure would be made safe.

case 10

TYPE OF FACILITY Hospital

TYPE OF PROBLEM Excessive settlement of foundations

Significant Factors

A. Inadequate depth of borings
B. Lack of inspection of subgrade before placing concrete for footings

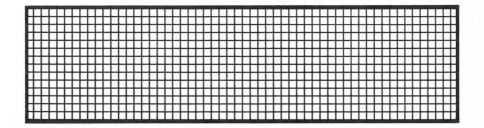

NARRATIVE

There are instances when cases of acknowledged defect in design and/or construction are resolved without the exchange of moneys, although the owner has incurred substantial costs for remedial work. Often, the resolution of such a dispute may be based upon real or imagined concerns known only to the owner. At other times, it is the desire to maintain a longstanding relationship that will prompt a negotiated settlement.

This case involves a ten-story addition to a hospital. Prior to the preparation of design drawings, the hospital had retained a soils engineer to perform a soils and foundation analysis of the project site. The resulting report was furnished by the owner to the architect. The architect, in turn, retained the services of a structural engineer and provided the report to him.

During construction, with most of the building erected, cracking of the masonry walls was observed. The cracking was traced to settlement of the columns in one corner of the building. Ten columns were affected. Settlement of up to 2 in was measured.

Remedial measures were taken, consisting of underpinning some of the ten columns and the injection of grout into the soil and rock underlying the rest.

The cost of the additional engineering analysis and of the remedial repairs totaled in excess of $400,000. Fortunately, the repair work proceeded smoothly and the opening of the facility was not delayed. This turned out to be a most favorable development since it avoided any subsequent claim by the hospital for loss of business profits.

The hospital thereafer commenced a lawsuit against the architect, the general contractor, the structural engineer, the original soils engineer, and a soils engineer who had been retained under the aegis of the general contractor to provide on-site inspections during construction of the foundation.

Interestingly enough, the hospital alleged that the soils engineer who prepared the original report was the agent of the architect. The complaint sought to hold the architect responsible for any errors in the soils report by virtue of the architect's failure to detect the actual conditions of the subsoil thereby permitting the settlement to occur during construction. The complaint also alleged that the structural engineer, as the agent and consultant of the architect, was negligent by his failure to require and obtain proper soils tests and analysis of the soils report, as well as a failure to properly inspect the pouring of footings by the general contractor.

It was alleged against the general contractor that he was negligent in failing to perform soils tests during placement of the footings, as well as in failing to properly construct the footings.

As additional elements of damage, the hospital contended that the repairs required to ensure the structural stability of the building precluded the future addition of two floors, as provided for in the master architectural drawings.

Findings prepared by the various experts who were retained appeared to place primary responsibility upon the initial soils report and upon the soils engineer retained at the behest of the owner to inspect the foundation construction. Additionally, the experts pointed out that the general contractor should bear substantial responsibility in that the footing for the column which showed the greatest settlement did not conform with the plans and specifications of the architect and structural engineer. Further, the contractor had not given notice to the inspecting soils engineer, nor to the architect, that he was about to pour certain of the column footings, thereby depriving a representative of the owner or the architect of the opportunity to observe the contractor's work for the purpose of ensuring compliance with specifications.

It was determined that the original soils engineer's report indicated a presumptive allowable bearing capacity of 20 tons/ft² which would have been sufficient for the design and construction of the building as planned. However, his report proved to be inaccurate as the building encountered severe settlement when a soil bearing pressure of less than 10 tons/ft² was imposed. It was the contention of the structural engineer that he had a right to rely upon the original soils engineer's report as to the adequacy of the soil bearing capacity.

Moreover, an argument was raised over the failure of the inspecting soils engineer to be present during the critical excavation and pouring of the footings. This same argument was also asserted in regard to the structural engineer. A review of the latter's contract with the architect indicated that the structural engineer was only required to perform on-site inspections at the request of the architect. No such requests were made by the architect. Thus it appeared that all parties would be able to point to the potential liability of the others at the time of trial.

To complicate matters further, three different soils engineering experts examined the settlement problem and disagreed as to the precise cause thereof. One expert deemed the problem to have been caused by deficient subsurface materials which should have been detected in the original soils report. The other two experts concluded that the settlement was caused by the existence of numerous voids and cavities. This latter argument enabled all parties to assert that these voids were unforeseeable and that responsibility for the settlement should therefore be borne by the hospital. It is to be noted that unanticipated soils conditions are generally a risk assumed by an owner if it can be shown that the soils condition encountered during construction could not have been detected by ordinary care and testing.

Shortly after the commencement of this action by the hospital, the architect filed a counterclaim seeking $300,000 for additional services rendered during the course of the project. Almost simultaneously, the general contractor filed a demand in arbitration for $360,000, alleging delay damage claims. It therefore became apparent that this matter would be vigorously litigated for many years and be further complicated by the separate court and arbitration proceedings.

It was at this juncture that the hospital communicated with the architect to determine whether a settlement could be arranged. The architect reiterated his long-standing commitment to work with the owner and to assist in any way possible to resolve this impasse. An agreement was thereafter executed between the hospital and the architect whereby both parties agreed to meet and discuss all aspects of the claim and to effect, if possible, a good faith settlement without prejudice to the ongoing litigation. All parties acknowledged that expensive and protracted litigation should be avoided.

As a result of these negotiations, a settlement agreement was finalized whereby the entire action was dismissed without an exchange of money. The architect agreed to a settlement of $35,000 for services previously rendered under its contract with the hospital and, in turn, agreed to render without cost, design services for the unfinished portion of the hospital. Mutual releases were also executed with the various other defendants and all parties expressed satisfaction that the matter had ended amicably.

TECHNICAL ANALYSIS

The hospital addition was supported, in spread bearing, on rock. As noted in the Narrative, design bearing capacity was 20 tons/ft^2 and the observed settlements occurred under a bearing pressure of only 10 tons/ft^2. The proximate cause was variously attributed to the presence of a layer of muck directly under the footings and to the presence of voids and cavities at a shallow depth below. This dispute never was resolved. The solution provided for both possibilities and consisted of underpinning some of the footings and grouting under the others. Thus, the technical problems were simple. Of interest is how they developed.

First, the soils engineer was selected on the basis of competitive proposals. Buying borings on the basis of bids is common practice, but not necessarily good practice. In some areas, the making of borings is treated as a professional service. At the least, architects and engineers must recognize the need to provide inspection of such borings; in some building codes such inspection is required by law.

Second, the contract for subsurface investigation required the borings to penetrate only to the bottom of the footings. Actually, since rock was encountered, the borings were penetrated 5 ft into rock, which, in some cases, terminated the borings *above* the level of the bottoms of the footings.

Normally, borings should penetrate to the base of the zone of influence under the footings. Since rock was encountered, there was some justification for lesser penetration. That the rock had cavities and voids was, perhaps, bad luck, or perhaps reflected a lack of experience with local conditions.

Third, neither the soils engineer's nor the structural engineer's contracts for services required them to inspect the subgrade for the footings. Instead, the owner retained, with payment through the general contractor, another engineer (the inspecting engineer) for this purpose, thus losing the advantages of continuity of services. To deprive architects or engineers of the opportunity to recheck and monitor their work is foolish. This applies to shop drawings as well as to field inspection. A "perfect" set of plans and specifications is more an ideal than a practicality. The checking of shop drawings and field inspection are necessary services. There is an advantage to awarding these services to the design team—it provides the opportunity for them to have one last review of their work. Certainly, having the inspecting engineer paid by the contractor is a well-known subject of debate. It is, in fact, illegal in certain jurisdictions.

Fourth, it developed that the inspecting engineer's representative wasn't present when the worst problem footing was placed. The architect had a full-time inspector on the job, who apparently also did not inspect the bottoms or advise the inspecting engineers that they were ready for inspection. Thus, the inspection which was provided did not recognize that the subgrade for the footings was a critical item of concern.

Finally, the contractor apparently never probed below the footings, as was required by his contract. But someone must have known that there was muck under at least one footing, because at the behest of the field personnel, the footing level had been dropped about 2 ft below the design level.

LEGAL ANALYSIS

Undoubtedly, had this matter not been amicably resolved, an enormous amount of time and expense would have been incurred in proceeding with the litigation. The claims by and between the various defendants, as well as the counterclaim against the owner for the acts of the original soils engineer, would have generated sufficiently complicated legal and factual questions to warrant a trial of several weeks.

Perhaps the concern on the part of the hospital that the architect and general contractor would prevail in their claims for unpaid services was real enough to obviate any desire to recover the $400,000 out-of-pocket costs incurred. There also may have been a recognition on the part of the hospital that the primary cause of the failure was the deficient soils report which had been relied upon by the structural engineer and contractor in pouring the footings. This would have been the responsibility of the owner, who had commissioned the soils report.

Whatever conjecture one might wish to apply to the possibilities in this case, it reflects one outcome among the many possibilities that may have been. To the extent that all parties expressed great satisfaction over the outcome of this case, the matter was successfully concluded. To the extent that design or construction errors were present so as to cause settlement of the building, it is the technical rather than the legal lessons which make this case history pertinent.

Comments

The lessons to be realized from this case are clear and of maximum importance, as the failings revealed by the narrative are of common occurrence. These lessons are as follows:

1. Too many parties to the design-build process can create opportunities for omission by default.
2. Continuity of service (preliminary design, final design, shop-drawing review, and site inspection) has value in promoting the detection of potential problems.
3. Borings need to penetrate *below* the bearing area under the footings and, preferably, to the base of the zone of influence.

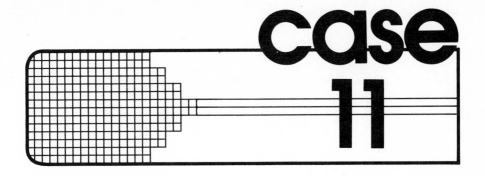

TYPE OF FACILITY Apartment house

TYPE OF PROBLEM Excessive deflection of floor slabs

Significant Factors
A. Drafting error
B. Failure to check deflections

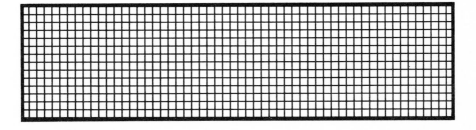

NARRATIVE

The structure involved in this case is a three-story apartment building. The first floor is a concrete flat slab over a basement. Above are three stories of wood framing, all supported by the concrete slab. Bay sizes were 20 ft by 29 ft.

During construction, and following placement of the dead load of the walls and partitions, the floor slabs showed excessive deflection. The amount of that deflection was a maximum of 2⅝ in at midspan and a minimum of 1³⁄₁₆ in.

The owner thereafter retained an independent consultant to review the structural adequacy of the apartment building. Tests were performed consisting of placing 55-gal water drums across the concrete slab, filling the drums to various levels and measuring the deflection of the slab. Applying the appropriate criteria established by the Uniform Building Code, and allowing a 24-hour recovery of the deflection after removal of the test load, it was determined by this expert that recovery equaled 71 percent of the maximum deflection. The Uniform Building Code criteria called for a 75 percent recovery. The expert, finding adequate explanation for the slight deficiency in the fact that the Uniform Building Code criteria were based on a slab with no superstructure on it, found that the slab had complied with the code requirements.

Notwithstanding the findings of this expert, litigation was subsequently commenced by the owner against the engineer (several years later). The action sought to recover $350,000, which represented the alleged cost of repairs and diminution in the market value of the building.

Further invesitgation concerning the adequacy of the design revealed the following proximate causes of the deflection:

1. The plans showed the bottom steel in midspan to be transposed. The column strip steel was shown in midstrip position and the midstrip steel was shown in the column strips.

2. The slab did not conform to the provisions of the Uniform Building Code with regard to slab thickness. A 9-in slab had been provided, whereas the code provisions called for a slab of approximately 11 in.

In his defense, the engineer argued that there may have been nondesign causes for the excess deflection. These included overloading of the slab before it was sufficiently cured, premature removal of shoring, failure to reshore, possible soil subsidence, and negligent construction by the contractor. Though there was some evidence to this effect, there was a general belief that such construction errors would form only a small part of the cause of the factors leading to the problem at hand.

Subsequent efforts were devoted to determining the true extent of the damages caused to the owner. These resulted in a determination that such damage was less than $150,000 if fully provable. Accordingly, negotiations were entered into and a settlement was reached with the owner in the sum of $95,000.

TECHNICAL ANALYSIS

The technicalities of this case are simple: the designer made an error. The lesson to be learned is that the structure must be checked for deflection as well as strength.

More interesting is that the structure was put into use and, so far as is known, remains in use—with its deficiencies and with knowledge that the deficiences existed. Remedial measures were limited to an epoxy leveling course on some floor panels and shimming the wood stud walls. The reason is that even gross deflections often do not affect function. The technical author of this text has pleasant recollection of an evening spent in the living room of an old house in Charleston, S.C., where if a ball had been dropped on the floor one would have been hard pressed to race it across the room to the fireplace. The weight of the fireplace and chimney had caused gross settlement of that side of the room; yet the home was lovely and eminently livable.

Most interesting of all, and the essential basis for continued use of the structure, was that the slab, despite the transposed reinforcement, was load tested and deemed to have satisfied the test requirements (Uniform Building Code). The transposition of the slab reinforcement apparently had little effect on overall (ultimate) strength—as a "collapse slab" analysis would tend to confirm. The test load was 0.74 (1.3 dead load + 1.7 live load) = 0.96D + 1.26L. The 0.74 factor was an adjustment of the normal criterion to reflect the fact that the design had been based on uniform loading over several contiguous bays, whereas the test load was applied only to one bay. Gross deflection was 1.45 in vs. 1.31 in, as allowed by the criterion $\Delta = L^2/20000t$. Recovery of deflection was 71 percent (75 percent required).

Another instance wherein a load test was used to prove that what appeared to be bad, was good!

LEGAL ANALYSIS

From a legal standpoint, one is immediately prompted to ask why the contractor was not brought into this action. At the time this action was brought, state law precluded a party alleged to be actively negligent from filing third-party claims against other actively negligent parties. Since the owner had chosen not to join parties other than the structural engineer, such a course of action was precluded as a matter of law. Limiting an action solely to one party, when others might arguably be liable, is a dangerous course of action for any owner. From a defense standpoint, it therefore becomes necessary not only to prove the engineer free of liability, but also to attempt to show other unnamed parties as the cause of the building failure.

The settlement effected in this matter represented an effort to minimize the ultimate damages which could have been recovered against the structural engineer. Moreover, it sought to avoid the extended cost of defending this action through trial. The structural engineer, after extended review of his plans by an independent expert, acknowledged several areas of design defect that were readily discernible. The wisest course of action, therefore, was to determine the precise amount of damages and negotiate a compromise settlement on the best possible terms.

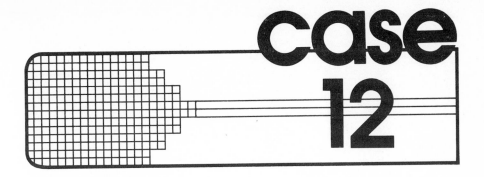

case 12

TYPE OF FACILITY	Prestressed concrete bridge
TYPE OF PROBLEM	Failure of bearings

Significant Factors

A. Insufficient travel in bearings

B. Bearings on adjacent spans locked

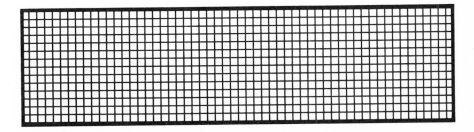

NARRATIVE

As part of an intrastate turnpike construction project, the plans called for the design and construction of a 1562-ft bridge with eight expansion joints, one on every other pier. The project was completed in 1965. Design of the bridge was undertaken by an engineer retained by a turnpike authority. Under separate contract, the authority contracted with a consulting engineer to undertake supervision of the various section engineers who prepared detailed plans. This consultant was responsible for planning, directing, and coordinating all shop drawings, mill inspection, and testing of workmanship and materials for conformance with the plans and specifications.

As part of the overall consulting contract, the consultant was required to inspect the entire turnpike system at least once a year, furnishing a report on the maintenance of the system and any recommendations necessary for repair. The authority agreed that if any of the yearly reports provided by the consultant suggested that the turnpike was not maintained in good repair, the authority would promptly restore the turnpike to satisfactory condition. Additionally, the authority agreed to employ, on its own staff, an experienced full-time chief engineer.

Shortly after completion of the project, the consulting engineer formed a joint venture with the designer who had acted as on-site resident engineer during the project. The purpose of the joint venture was to permit the resident engineer to undertake full responsibility for the ongoing annual inspections. Approximately 1 year later, in 1967, the joint venture was dissolved and the consultant turned over all obligations under his contract with the turnpike authority to the resident engineer and discharged fully his further obligations thereto, taking back the "hold-harmless" from the resident engineer for any future liability.

In January 1976, a maintenance supervisor for the turnpike authority noted that an expansion joint on one of the piers of the bridge in question had dropped 1½ in. The bridge had been inspected annually since the turnpike was open to traffic in 1966. From all prior annual reports, it appears that the entire bridge structure and expansion joints had apparently functioned properly from that date until January 1976.

In his annual inspection report to the turnpike authority, the resident inspecting engineer described the failure and the subsequent developments as follows:

> On January 6, 1976, it was reported that the bridge, north-bound lane, had dropped approximately 1½" over pier No. 7 causing the prestressed beams to start to slip off the sliding bearings and rocker plates. This occurred after a temperature change of approximately 70° in 36 hours. There are 8 expansion joints in this structure with Pier No. 7 being approximately in the center.

It is believed, at this time, that the other expansion joints acted abnormally causing the majority of the contraction to be concentrated at this joint for some unknown reason. This expansion joint is approximately in the center of the bridge. Inspections of this bridge were continued and movement of the beams began putting undue lateral pressure on the pier cap causing the pier to be in a state of collapse and the Turnpike Authority was forced to close this section of the turnpike on January 15 until such time as the bridge could be placed in a safe operating condition. Subsequently bids were taken for removal and replacement of this pier and the method decided upon was placing structural steel bents on a concrete footing, jacking up the spans to grade, thereby allowing the bridge to be opened to traffic. This was done and this section of turnpike reopened to traffic on June 15, 1976. In an effort to do all possible to prevent a reoccurrence of this problem during reconstruction of the new pier, all bronze expansion plates and steel rockers were replaced at the other expansion joints as well as this expansion joint with Neoprene expansion assemblies.

As a result of the 6 months of repair work, the turnpike authority incurred a loss of $257,500 for lost revenue. The insurance company which paid this sum to the turnpike authority commenced an action to recover its payment. This action was brought against the design engineer for the bridge, the consulting engineer initially responsible for inspections, the resident engineer who assumed the responsibility for inspections subsequent to dissolution of the joint venture, and the contractor on the project. Initial investigaton by the consulting engineer and the resident engineer failed to indicate any apparent cause for the 1½-in drop in the bridge structure. Assuming proper construction procedures, maintenance and normal weather conditions, the structure was deemed to have been safe for vehicular traffic for a period of between 45 and 60 years before superstructure maintenance would be required. Annual inspection reports by the resident engineer failed to indicate anything other than a normal wear and tear.

A review was undertaken of the insurance policy which plaintiff had issued to the turnpike authority. That policy contained an exclusion for "inherent defect, wear and tear, gradual deterioration, or expansion and contraction due to changes of temperature and loss resulting in collapse." Thus the question arose as to whether or not the 1½-in drop in the structure would be deemed a "collapse" and whether or not the deterioration was due either to an inherent defect or to wear and tear. If the attorneys could show that the insurance company had failed to invoke an exclusion and improperly paid the turnpike authority, the suit could be dismissed as a matter of law.

Attention was also drawn to the fact that the alleged collapse occurred more than 10 years after completion of the bridge. This raises the question of whether a court would rule that the statute of limitations barred the maintenance of this action against the design professionals.

On behalf of the consulting engineer, it was argued that the design of the bridge was undertaken by the original engineer of record, that the consulting engineer had prepared only one annual inspection report before those duties were assumed by the resident engineer for the next several years, and that the turnpike authority bore responsibility for maintenance and repair of any conditions which required such attention.

After initial depositions of the parties were taken, the defendants filed motions for summary judgment seeking to bar the action brought by the insurer. The motions alleged the applicable statute of limitations as well as the fact that the insurer's action in subrogation (i.e., in its own name on behalf of its insured, the turnpike authority) was likewise improper, and sought the dismissal of the action before trial. Shortly after these motions were filed, the insurance company, deeming there to be some merit to the motions, commenced settlement negotiations. A settlement in the sum of $30,000 was achieved, in which the consultant paid $13,500, the resident engineer, $12,500, and the original designer, $3000.

TECHNICAL ANALYSIS

The structure under consideration is a highway bridge constructed using prestressed concrete beams on a concrete substructure. The construction is shown schematically in Figure 12-1. The overall length of the bridge was 1562 ft, with spans averaging about 110 ft. There were eight expansion joints in the bridge, located at every other pier, as indicated.

As noted in the Narrative section, almost exactly 10 years after the bridge was opened to traffic, and during the month of January, the maintenance supervisor reported that the northbound roadway of the bridge had dropped 1½ in at one of the expansion joints near the center of a series of spans. The event followed an abnormality in the weather wherein the temperature had dropped from 73°F to 3°F (i.e., 70°F) in a period of 36 hours.

What happened was that the expansion joint at this particular pier had opened to the extent that the prestressed beams had walked off the bearing and dropped onto the pier cap. A detail of the expansion bearings used in the structure is presented in Figure 12-2. Investigation revealed that

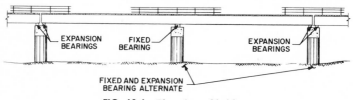

FIG. 12-1 Elevation of bridge.

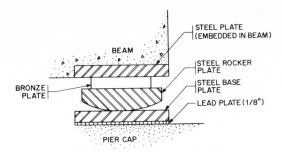

FIG. 12-2 Detail of expansion bearing.

the other expansion joints in the structure had not moved as planned. The majority of the contraction in the entire series of spans had accumulated and developed at the one joint.

In the succeeding weeks, further movement developed at this joint (in both northbound and southbound roadways). Examination of the pier (which is shown schematically in Figure 12-3) indicated that the unbalanced lateral loads which had developed incident to the occurrence had broken the 4-ft-diameter sections of the columns of the pier. The bridge was closed to traffic.

Repairs consisted of lifting the deck and rebuilding the broken portions of the pier. More significantly, the owner decided to, and did, replace *all* of the expansion bearings in the bridge with elastomeric (neoprene) pads,

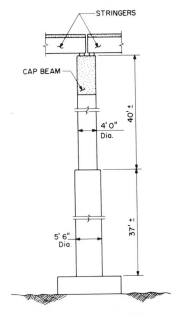

FIG. 12-3 Elevation of pier.

the use of elastomeric pads being the standard detail in use for the bearings on bridges of such spans and construction at the time of repair (i.e., the standard detail had changed in the 10 years that the bridge had been in service). The bridge was closed for 6 months to accomplish these repairs.

Comment

This case is included because it illustrates a pattern familiar to those involved in the maintenance of bridges. Aside from the deck, the bearings and the expansion joints tend to be the biggest maintenance problem. Locked bearings and the accumulation of movement in a single joint are a common occurrence.

LEGAL ANALYSIS

An astute reader of this case will raise several questions. First, one might question the propriety of the insurance company for bringing this action in the face of an existing statute of limitations which clearly appeared to bar a cause of action for negligence against the defendants. Second, one is led to wonder why the consulting engineer did not immediately tender his defense to the resident engineer who had assumed all obligations under the joint venture inspection agreement and furnished the consultant with a hold-harmless agreement designed specifically for such a lawsuit. Finally, one is led to inquire into the reason for a settlement of $30,000 before a decision was handed down by the court on the motions for summary judgment.

In regard to the propriety of the lawsuit, it is quite clear that the insurance company was aware of the applicable statute of limitations which appeared to bar this action in the first instance. However, the insurer argued that the original consulting agreement created a breach of an implied warranty on the part of the defendants that the completed bridge would be reasonably suited for its intended purpose and would last and serve that purpose for a reasonable length of time. Accordingly, the insurance company cited the testimony of the consulting engineer, who stated that the bridge, had it been properly designed and constructed, would have continued to function without incident for between 40 and 60 years before superstructure maintenance repair would have been required. On these grounds, the insurer argued that the bridge, as constructed, was not reasonably suited for its intended purpose and that it did not serve such purpose for a reasonable length of time as per the contract provisions. By seeking to assert such a breach of warranty, the insurer argued that the breach did not manifest itself until the date of the discovery, i.e., January 1976, and hence the statute of limitations did not begin to run until that date and the action was timely. The insurer sought to have the court view this case as one sounding in contract and not in negligence. Were this strategy successful, the action would have proceeded against the defendants in accordance with their contract obligations, rather than the duty of care they owed to the turnpike authority.

The second question raised involves the legal intricacies of a hold-harmless agreement. There is a large body of law which has construed such hold-harmless provisions, upholding the validity of these agreements in some instances and declaring them void as a matter of public policy in others. For purposes of this casebook, it is not necessary to discuss the legal validity of the hold-harmless language which was included in the agreement dissolving the joint venture between the consulting engineer and its resident. What is important is that architects and engineers who are either relieved from future obligations on a project, or voluntarily assign those obligations to another, must have agreements to this effect reviewed by their attorneys to ensure that all existing law on the subject is incorporated into the language utilized. A careful drafting of such provisions will make the difference between defending a lawsuit or being able to tender the defense over to the party who purportedly intended to assume the obligation for any future liabilities arising therefrom. Any ambiguity will generally lead a court to defer the application of such a hold-harmless agreement until the time of trial. The additional cost involved, and the potential liability which may be faced by the party who thought it was free and clear from such exposure, is not worth the risk of signing a loosely drawn agreement to this effect.

In this instance, the resident engineer who performed the inspections after the joint venture dissolved had died before the commencement of the litigation. As a result, his estate was also joined as a party defendant. Had the litigation proceeded further, undoubtedly a cross claim for indemnification, citing the hold-harmless wording, would have been filed. A motion to compel the resident to assume the defense and hold the consultant harmless from any liability would likewise have followed.

The third question involved the propriety of settling this matter prior to a ruling on the motion for summary judgment. In almost every state, there has been a substantial body of case law in the last few years construing the applicable statutes of limitations relating to the design professions and construction industries. More than a dozen state courts have declared unconstitutional specific statutes of limitations passed by the legislatures and designed to bar actions brought more than a specified number of years after completion of design and/or construction. The constitutionality of these statutes, where they exist, will undoubtedly continue to be a fertile source of controversy for many years to come. Asserting such a statute

as an affirmative defense at the outset of litigation against a design professional is imperative. Familiarity with the leading case law in each state
and around the country is also necessary for counsel defending an architect
and engineer.

Here, the plaintiff sought to distinguish between the negligence statute
of limitations, which barred a cause of action after a specified number of
years, and contract statute of limitations, in which the period of time commenced to run after the alleged breach occurred. This was a plausible
theory; the court could have deemed an action of this kind involving substantial damages, to warrant a denial of the motions for summary judgment,
thereby permitting the matter to proceed toward trial. On the other hand,
the insurer, sensing that there was merit to the motions, decided that a
negotiated settlement would be in its best interests in lieu of an outright
dismissal and the subsequent costs of a possible appeal.

For this reason, the settlement became palatable to all parties, and the
contributions of each essentially represented the future cost of litigating
this matter through to the time of trial.

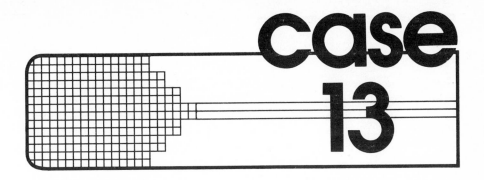

case 13

TYPE OF FACILITY   Sewage treatment plant

TYPE OF PROBLEM   Strength of posttensioned, concrete,
                        rigid frames

Significant Factors

A. Design philosophy.

B. Yield capability of connections in posttensioned construction.

C. Is a design which requires yield of the structure for the development of adequate strength an acceptable, functional entity?

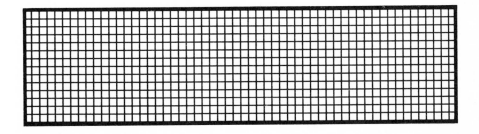

NARRATIVE

The structure involved in this case is a building included as part of a municipal sewage treatment plant. Construction consisted of precast, prestressed concrete frames, with posttensioning used to develop continuity in the frames. The problem concerned the strength of the precast girders.

Two types of girders were involved. Type 1 supported a lesser loading and type 2, a heavier loading. During the shop-drawing stage, the detailer detected an apparent deficiency in the design of the ends of the type 1 girders. Evidently, he was correct; the structural designer prepared corrective drawings and the corrections were made prior to fabrication of the girders. As a result, however, the prime consultant expressed concern about the design of the type 2 girders and commissioned a third consultant to review the entire structure. This consultant found that the type 2 girders had been designed for end moments of about 70 ft-kips, whereas the actual moments expected to develop were 480 ft-kips. The end connection of the girders and the ends of the girders themselves had no hope of developing this resistance (see Figure 13-2). Accordingly, some of the type 2 girders

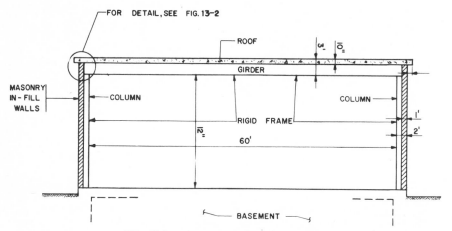

FIG. 13-1 Schematic section of building.

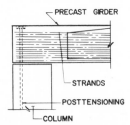

FIG. 13-2 Detail of girder-column connection.

were replaced[1] at added cost, with new, stronger girders (type 2A). Contractor claims in excess of $150,000 were subsequently filed with the owner, who in turn sought recovery from the engineer. Through negotiations, these alleged damages were reduced when the subcontractors agreed to accept $68,000 in settlement of their claims.

In addition to negotiations between the architect and the subcontractors, separate negotiations were held between the structural engineer, the architect, and the owner, regarding the manner and apportionment of any settlement of the claim. On behalf of the structural engineer it was argued that the revised design represented a betterment to the municipality which should be absorbed by the owner. This figure was placed at $18,000. Accordingly, since the owner would have had to pay in the first instance for such original cost had the facility included the "improved" design concept, the owner was thereupon asked to participate in the additional costs required for the repair work. Additionally, the architect was advised that his participation would be necessary for any substantial negotiations to proceed. This argument, of course, had little effect upon the architect, whose own experts had verified that the defective condition resulted from errors solely within the realm of the structural engineer.

Curiously, although negotiations were concluded with the subcontractors for settlement of $68,000, and although the architect and owner had incurred additional expenses in the remedial repair work, no action was ever started in this matter nor was any payment ever made by the structural engineer.

TECHNICAL ANALYSIS

On the surface, the Narrative of this case presents a simple case of design error—but not so! The structural designer contended that while he may have underestimated the negative moments at the knees of the frame, he had designed the girders for the correct total moment (positive *plus* negative—the simple beam moment). He argued that the girders consequently had ample load capacity and that all that would have happened as a result of the insufficiency in end-moment capacity would have been some cracking and gapping of the end joint—but that the structure would have supported the loads. Lateral stability was to have been provided by the infill masonry walls at the ends of the building and by the diaphragm action of the roof.

This is a most interesting contention and one to which the consultant reviewers appear not to have responded directly. The correspondence and reports by the consultants refer to ACI-318-71 provisions relating to permissible redistribution of negative moments in flexural members (based

---

[1] Some were used in place of type 1 girders

on the code in force at the time, a maximum 10 percent redistribution was permitted) and that such provisions do not relate to prestressed members. Clearly, the design was in violation of ACI-318, but would the frames actually have collapsed? Possibly not!

The attorneys for the structural engineer proposed to the plaintiff that a load test be conducted on one of the discarded girders, on a "loser pay all" basis, with the ends of the girder being simply supported. The purpose was to prove or disprove the engineer's contention. The plaintiff declined, giving the following reasons:

1. The construction schedule did not permit such testing prior to the decision to discard some of the girders.
2. The results of the test might not be conclusive.
3. Several tests would be required to simulate different loading conditions.
4. Strength alone is not a criterion for acceptance.

The development of cracking at the ends of the girders due to the deficiency of the end connection would be objectionable, and the load test never was performed. Too bad! It would have been interesting.

LEGAL ANALYSIS

This case is an anachronism in that it represents one of the few instances in which damages were incurred but, for one business reason or another, no affirmative action was taken to recover the loss sustained. Counsel will encounter situations such as this from time to time. Although they are not easy to explain, various hypotheses can be suggested for such a lack of the characteristic zeal typical of an owner in a situation such as this.

Quite often, local governmental projects are funded, in whole or in part, by federal or state funds which permit a certain degree of latitude in regard to construction costs. Informally, many state and federal projects do include percentage guidelines for contingencies, sometimes in an amount of up to 5 percent of the cost of construction. These guidelines are not intended to condone design or construction deficiencies; rather, they represent the exigencies of the construction process and the fact that change orders and other departures from the original design concept will be required during the scope of any project.

A second possibility concerns the cost-effectiveness of commencing an action to recover losses of less than $100,000. In this day of sophisticated litigation, commencing an action involving technical engineering problems requires the use of sophisticated counsel and independent experts, the cost of which very often can militate against a recovery which would offset the cost of prosecuting the action. While this may appear as alarming to some, those who are knowledgeable in the field of construction and design litigation are aware that there is no shortcut to proceeding with this type of action except in the most clear-cut of issues and with the barest minimum of parties involved.

Of course, there are some who would argue that the answer to the high costs of litigation rests with the need to invoke the arbitration process in all such disputes. The authors are clearly of the opinion that arbitration is not the easy answer for the design profession when it comes to resolving problems of the kind described in this book. While this is not the forum to set forth a definitive argument against arbitration as the answer for all such disputes, it should be pointed out that the arbitration process no longer is the expedient, cost-conscious method of resolving construction disputes that it once was.

Finally, one last theory should be mentioned as a possible reason why litigation was not commenced by either the owner or the architect against the structural engineer. The possibility exists that a review of the files of the owner or the architect disclosed knowledge on their part that problems existed with the structural design. If such knowledge did exist, the failure to act upon learning of the design defects at an earlier time would place a portion of liability upon either the owner or the architect, thus reducing any damages that might be recovered against the engineer. Consequently, one or more of the above factors may have played a significant part in precluding the start of litigation in this matter.

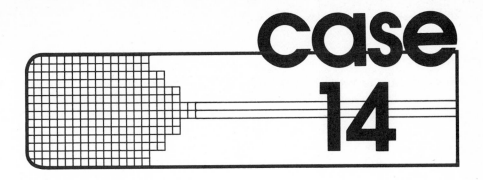

case 14

<table>
<tr><td>**TYPE OF FACILITY**</td><td>Industrial building</td></tr>
<tr><td>**TYPE OF PROBLEM**</td><td>Collapse of roof in heavy rain storm</td></tr>
</table>

Significant Factors

A. Storm intensity exceeding 25-year recurrence interval used for design
B. Controlled-flow drainage system for roof
C. Architectural details not consistent with implications of controlled-flow drainage system
D. Tenant changes

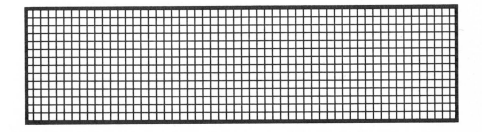

NARRATIVE

The structure involved in this case is a single-story, steel-frame (trusses) industrial building used for manufacturing. Schematically, the structure is shown in Figures 14-1 (Plan) and 14-2 (Section).

During the course of a major storm, consisting of heavy rainfall and hail, during which time between 7 and 11 in of rain fell in a 2- to 3-hour period, two areas of the building's roof collapsed. A total of 40,000 ft² was affected. The storm was later rated as varying from a 50-year storm to a more than 300-year storm. This means that according to the weather history of the area in question, a storm of this severity would be statistically expected to occur once every 50 through 300 years. Without question, the storm exceeded the design criteria for weather conditions in that locale. Locally, the storm caused widespread flooding, 4 deaths, 70 injuries and approximately $16 million in damage. With respect to the building in question, losses exceeded $3 million.

The architect had been retained by the owners to prepare preliminary and final construction plans and specifications. The architect also furnished a field representative during construction who performed periodic inspections under the supervision of a resident engineer employed by the owner. The general contractor performed site preparation work and also constructed the structural elements of the building.

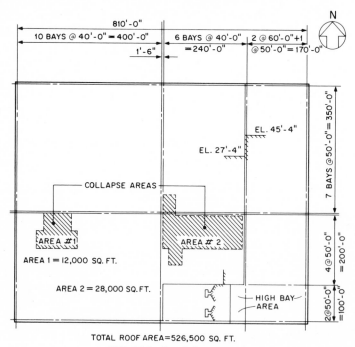

FIG. 14-1 Roof plan.

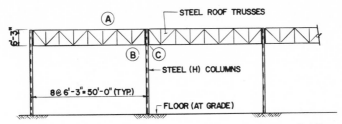

FIG. 14-2 Schematic of typical section of roof. (Roofing and siding, purlins, and girts not shown.)

An interesting feature of this project involved the intensive participation of the owner in the preparation and review of the design for the project. The owner maintained a complete engineering staff which coordinated the design and construction and was intimately involved in the selection of all design criteria. In fact, the owner furnished its own guide for design and construction of the project and required the architect to stay within the criteria established by the guide. All plans and specifications were reviewed by the owner's engineering staff and the owner's engineers were required to review any changes in the design which were required during the construction of the building.

Following the collapse, a lawsuit was started by the insurer for the owner to recover the repair costs incurred in restoring the building to its original condition. Interestingly, the owner did not join in this lawsuit and at no time asserted a claim for the costs and damage it had sustained as a result of the collapse. Hence, this action involved a claim by the insurer for the losses it had sustained in payments to the owner. It was claimed that this amount was $1.6 million. The lawsuit named as defendants the architect and the contractor.

Both the architect and the general contractor asserted that the severity of the storm represented an "act of God" and asserted this defense in the litigation. Certainly, the weather records of the locale indicated that there was no prior history of a storm of such intensity. Nor did the local building code establish a criterion for the structure to meet a storm of this type.

A team of experts was retained on behalf of the architect to analyze the structural design. Questions were raised about the adequacy of roof drainage, evidence of contractor error in the placement of structural columns, improper roof maintenance by the owner, and a potential design error by the architect which caused depressions of up to 2⅝ in. in the roof prior to the storm. Analysis was also undertaken to determine whether a correlation existed between the wind direction and the areas of the roof which collapsed. It was surmised that the wind may have driven the ponding water against a high wall of one of the roof areas, causing excessive ponding.

Consideration was also given to the possibility that the wind may have created a suction or slapping effect in which the water was suddenly picked up and violently dropped against the roof. Initially, it was determined by the experts that the weight of the water itself would not have been sufficient to cause the collapse, as the strength of the structure and the roof had sufficient design capacity to withstand such accumulations.

While the experts were able to establish that the architect's original design conformed to the engineering requirements established in the applicable codes and in the owner's design guide, a question was raised in regard to certain mechanical equipment which had been added by the owner and hung from columns and trusses within the building. The owner also appeared to have removed certain diagonal truss members in order to install this equipment. Photographs of certain columns within the building showed that they had incurred damage from repeated collisions by forklift trucks which were used in the building.

During the early stages of this litigation, the parties reached a settlement of the matter in the sum of $500,000. The architect contributed $350,000 and the contractor, $150,000. The owner at no time asserted a claim against the parties for the amount of loss it had sustained and no reason was provided for this position, notwithstanding its full knowledge of the litigation.

TECHNICAL ANALYSIS

One of the questions raised in this case was the propriety of the designer detailing the structure to support a roof load (ponded water) corresponding to a storm of a 25-year recurrence interval. This corresponded to a design live load (actually used) of 20 lb/ft². This value had been prescribed by the owner and reflected the fact that the building was in an area where no substantive snow load was to be expected. The designer accepted this value, and eleven different parties (the experts who reviewed the design after the collapse occurred) agreed that it was a reasonable value. The complication was that the roof was designed for controlled-flow drainage, i.e., the roof drains were not sized to convey the runoff as fast as it collected, but for a lesser capacity so that water would pond on the roof during peak rainfall and flow off the roof as the peaks of the storm passed. This resulted in a saving in the sizes of the downpipes, the sewer pipes, and the entire site drainage system. The significance of the period of recurrence (the design storm) is that the design live load corresponds to a depth of ponding of about 3½ to 4 in and the drain system had to be sized so that the depth of ponding would not exceed this value. In fact, the roof was designed with the intention of ponding about 3 in of water.

It is conventional to design a roof for a 25-year storm. The snow charts in most codes are for a 25-year period of recurrence. The structural design manual used by the U.S. Navy (NAVFAC DM-2.2) specifically calls for structures to be designed for a 25-year service life and gives design loadings based on a 25-year period of recurrence. However, the facts are that we normally use our structures for longer than 25 years and that we do not want, or expect, them to collapse or even to be damaged by a storm of 50- to 100-year recurrence. How to explain this apparent paradox? The explanation is simple. We trade on the factor of safety. A roof, such as the one in this case, which is designed for 20 lb/ft² live load and about 15 lb/ft² dead load should support, at least, 0.4* × 15 + 1.7 × 20 = 40 lb/ft² of imposed load—all other things being ideal. In this particular case, because many of the truss members were proportioned to conform to "minimum-size" requirements, the calculated live load capacity was 27 lb/ft², not 20 lb/ft², giving an ultimate capacity of about 45 to 50 lb/ft², or about 10 in of ponding. Why, then, did this roof collapse?

As might be expected, the answer is complex. In essence, it is because all other things were not ideal. There were imperfections in construction. The owner had made modifications which increased the imposed dead load and which impaired the structural capacity of some of the trusses. Columns had been damaged by impact from forklift trucks. All of these things used up a part of the safety factor so that when the depth of ponding exceeded the design value of 3 in (which appears almost certainly to have occurred) there was not enough safety factor left to support the imposed load.[1]

Imperfections in Construction

1. Poor welding of roof deck to its support reduced the lateral bracing afforded the purlins and trusses.

---

* The normal load factor (1.4) less 1.0 for the fact that the roof is supporting its own weight.

[1] This raises the philosophical question of who the factor of safety is intended to protect. Designers often consider it to be for their exclusive benefit. But this is not so. The safety factor is for the protection of everyone involved in the construction *and use* of the structure. Designers need protection against inaccuracies in the loads which they must estimate (wind, snow, seismic effects, waves; even the dead load) and against variations in the properties of the construction materials. Contractors also need the safety factor for their purposes. They must store materials in a handy location (see the stacks of brick and block pallets in a building under construction). They must have the capacity to support their construction machinery (concrete buggies, for example, or erection cranes) and to accommodate the vagaries of workmanship. They assume that the normal safety factor exists. If not, if they have to restrict their operations because the design is unusually sensitive, their costs are affected and they are entitled to know that the design is unusually sensitive. The owner is entitled to some flexibility in the use of the structure. It is common to find local "overload" in an office caused by file cabinets and storage, or a computer, or a small safe. If something occurs outside the building and everyone runs to the window to see, the design live load is likely to be exceeded, locally. Some lack of maintenance must be expected. All this is part of normal usage. Fault lies not with the fact that any one party's performance is deficient, but with the party or parties who use more than their share of the safety factor. This is also the reason that when a collapse does occur, it is usually the result of a compounding of several deficiencies, not a single cause.

2. Camber of roof trusses may have been omitted. Roof was built with sagged areas ("hollows").

Owner Abuses

1. One of the diagonal web members in several of the trusses had been removed to provide clearance to accommodate the installation of ducts.
2. Process equipment and piping was hung from the roof. An allowance of 3 lb/ft² was used in the design. Actual loads as much as 13 to 15 lb/ft² were found.
3. Catwalks had been added on the roof, and additional bridge cranes, not contemplated in the design, had been installed.
4. Columns had been damaged by impact from forklift trucks.

Design Deficiencies

1. The various reports developed relating to the failure of the roof drainage system concluded that no controlled-flow drainage system can cope with a high-intensity rainfall of protracted duration. *Relief must be provided by scuppers or similar overflow devices.* Adequate overflow devices were not provided in all areas of the roof.[2] For example, Area 2 was confined by a perimeter expansion joint projecting 7 in above the roof. Accordingly, 7 in of ponding was required in the area before overflow occurred.
2. The roof trusses reputedly were designed to have a camber of ⅛ in. in 10 ft. There is a question whether the camber was provided, and, if not, why not? In any event, the roof, as built, was not properly graded to drain. There were low points, or "hollows," up to 2½ in deep which caused additional depth of ponded water (in the amount of 13 lb/ft²).
3. If we add the effects of the above (7 in ponding to cause overflow, plus 2½ in of ponding because of sags in the roof) total ponded load is almost 50 lb/ft² or, as indicated above, enough to use up all the design safety factor and bring the structure near to collapse, even if there were no other deficiencies. In Area 2 it is possible that this was the major factor. In Area 1, the area was confined by a joint of lower profile so that overflow occurred under a lesser depth of ponding. Here, some or all of the other factors listed had to be of significance.

[2] These reports also concluded that the flow capacity of a roof drainage system designed for a high-intensity storm tends to be limited by the capacity of the runs of piping between the roof drains and between the roof drains and the sewer, rather than by the roof drains themselves. For this case, in order to reduce storage during a high-intensity, protracted storm to 3 in, almost as much drainage was needed as for 100 percent immediate runoff. The controlled-flow roof drainage system was found to be largely ineffective for a high-intensity storm.

4. An error was discovered in the design calculations. The properties of the truss had been entered into the computer improperly, i.e., the top chord substituted for the bottom chord and vice versa. This proved to be of little significance, but demonstrates the need for care in using the computer.

5. One of the experts who reviewed this case noted that the structure (see Figure 14-2) had been designed as a semirigid frame, i.e., simply supported for dead and live loads, but having a capacity for the development of moment at the connection of trusses to columns to resist lateral wind forces. He pointed out that this unintentionally caused compression in the bottom chord member *BC* (see Figure 14-2) under dead and live load and this compression was sufficient to cause member *BC* to buckle. Other experts argued that such buckling merely caused the truss to revert to simply supported behavior and that since the observed problem clearly was related to vertical, and not lateral, load, the buckling, if it in fact did occur, would not reduce the capacity of the truss to resist vertical loads. Semirigid construction [American Institute of Steel Construction (AISC) design specification, type 2] is in common use. It is desirable to check that the connection which is assumed to be semirigid can rotate without fracture or buckling. Note the similarity of this problem to that of Case 13, where the adequacy of the rigid frame depended on the yield of the end connections.

Remedial Measures

1. Lower profile for roof expansion joints (see Figure 14-3). This was done to reduce the depth of ponding which could occur before overflow.

2. More roof drains.

3. Tie the columns together at the expansion joints (see Figure 14-4).

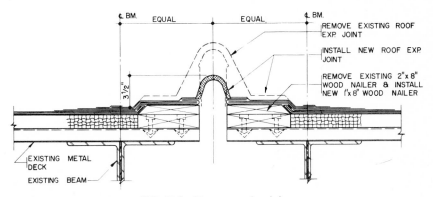

FIG. 14-3 New expansion joint.

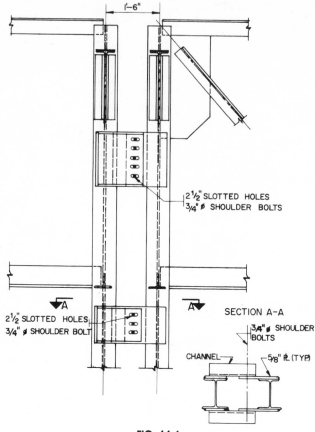

FIG. 14-4

This was done to improve stability and, perhaps, to relieve the compression in the end of the bottom chord of the trusses.

4. Some member sizes were increased to accommodate the additional loads due to tenant changes.

5. Miscellaneous other improvements to roof drainage system.

Comment

The responsibility of an owner in the development of a design and in the use of a structure bear noting. For example, if a low level of maintenance will be provided, the owner has no right to the benefits of an interior roof drainage system which depends for its function and safety on blocked or clogged drains being promptly detected and repaired. It is the owner's responsibility to proclaim his or her intentions to the designer concerning the following:

1. Possible changes in occupancy or use
2. Expected level of maintenance of nonfunctioning entities (painting, pointing of masonry, repairs to roofing, etc.)
3. Expected level of maintenance of functioning entities (roof drains, mechanical ventilation for enclosed spaces, etc.)

A second matter of note is the design of a controlled-drainage system for a roof. This design must consider the following:

1. Level of roof as influenced by camber, elastic deflection, creep and other inelastic deflection, and construction tolerance
2. Rate of runoff as influenced by size and slope of entire drainage system, including the final outlet into the sewer
3. Possibility of clogged drains or pipes
4. Overflow when ponding reaches critical depth

LEGAL ANALYSIS

From the standpoint of the attorneys representing the architect, this case was fraught with pitfalls and problems throughout. In the first instance, there was no question but that disaster had struck the building. Damages of several million dollars had been incurred, and, absent a determination that the failure was attributable to an "act of God," it would be an exposure to either the architect or the general contractor. For reasons that were never clarified, the owner chose not to assert any claim against the defendants, although it incurred uninsured damages of $2 million. This created an element of uncertainty, since the owner could have chosen at any time to sue for its losses.

A team of attorneys was assigned to coordinate the retention of outside experts, review the applicable building codes, interview prospective witnesses, research available defenses and review the enormous amount of documentation surrounding the project. It became quite clear at an early stage in the litigation that a viable defense would be the contention that the architect complied fully with the design criteria established by the owner, since the design was fully approved at all stages and accepted by the owner's engineering staff. However, elements of this defense posed a threat that such a challenge to the owner would prompt the filing of an action by the owner to recover its losses. Hence, there was a need to proceed cautiously with such a defense against the insurer for fear that it would substantially increase the exposure to the architect.

At all times, a strong indication was given to the insurer and the general contractor that a vigorous defense on several levels would be asserted by the architect. By projecting such a strong defense effort, the insurer was prompted to reconsider its decision to prosecute this action vigorously. Consequently, settlement negotiations did prove to be fruitful in the face of protracted litigation and the extensive costs which could have been anticipated for expert and legal fees.

The possibility of an early settlement was perceived shortly after the litigation began. Overtures were made by counsel for the architect to the attorney for the plaintiff insurer to determine whether the insurer had undertaken a detailed analysis of the exposure to the defendants. Counsel for the insurer indicated that, while he believed design and/or construction errors could be proven, lines of communication for settlement should be kept open by the parties. This response signaled an indication that no hardened position would be asserted by the insurer as the matter proceeded. It also held out the promise, ultimately fulfilled, that settlement negotiations on an amicable basis would be the most fruitful way to proceed.

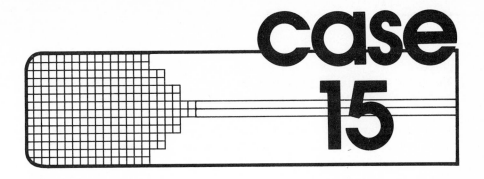

case 15

TYPE OF FACILITY

Apartment buildings

TYPE OF PROBLEM

1. Buckling of columns
2. Sag of ceiling
3. Settlement of floor slab
4. Tilting of foundation walls

Significant Factors

A. Marginal quality of the plans
B. Poor estimation of loads
C. Lack of compaction of subgrade for slab on grade
D. Improper construction sequence

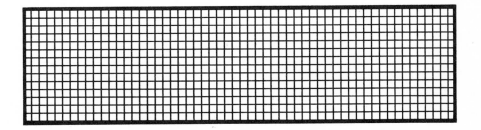

NARRATIVE

One of the most difficult moments in the life of design professionals must occur when they are confronted with the existence of a substantial failure on a project they have designed. Faced with the prospect that a failure has occurred, and that the owner has, or will, incur substantial costs for remedial repair work, an initial reaction of despair is not uncommon.

This next case involved a situation where at the outset even the attorneys for the architect were of the belief that substantial liability would result, only to find that by pursuing the case diligently, a favorable, almost nominal, settlement of the action was effected.

The project involved the design and construction of an apartment complex comprising two four-story, 42-unit buildings, a five-story, 51-unit building, and a one-story clubhouse. The buildings were wood frame, with bearing walls and Lally (concrete-filled pipe) columns. Plans, including all structural designs, were prepared by the project architect. Under his agreement with the owner, the architect had no responsibility for inspection obligations during construction.

The pleadings ultimately filed in this action against the architect alleged that the plans and specifications were so negligently and carelessly prepared that columns in certain parts of the buildings began to crack, buckle, and give way, requiring the cessation of construction, shoring of columns, and installation of heavy-duty columns to replace those originally specified by the architect. It was further contended that a wall began to tilt during construction, causing the owner to halt work to reinforce and correct latent deficiencies. In connection with the clubhouse, it was asserted that the architect had failed to specify foundations or footings under the block walls of a shower room, causing the floor to sink and crack and the block walls to fall. Damages of $500,000 were sought against the architect.

Specifically, the owner contended that the plans provided by the architect for the four-story apartment building were defective in that they called for 3½-in Lally columns, which was deemed by the ironwork suppliers to mean lightweight Lally columns. Thus, lightweight columns were installed and in place during construction. It was a local building inspector who noted that these columns appeared to be undersized and recommended that heavier weight (same diameter, but greater wall thickness) columns be installed. As a result of this removal and replacement, the owner contended that a 4½- to 5-month delay in occupancy of the buildings occurred.

With respect to the design, the owner asserted that the architect's plans were defective in that they failed to provide for a joist of sufficient strength to support a 20-ft span at the point where the entrance foyer met the first-floor corridor, and that the same defect affected each of the floors above. It was argued that the insufficient strength of the joist caused the building to sag in the middle, threw apartment doorways out of plumb, caused cracks and openings in the drywall, and also resulted in uneven floors.

Similar claims of design defect were asserted in regard to several of the other buildings. Additionally, it was contended that one of the buildings had begun to list, causing a Lally column to bow and supporting concrete walls in a portion of the building to lean dangerously. It was contended that an outside expert was called in, following which steel plates were welded to the sides of the Lally columns and an additional wall was installed to retain the shifting wall.

In connection with the clubhouse, it was asserted that the plans were inadequate in that the 4-in concrete floor which was poured in the basement was not strong enough to support the inner cinder block walls. Consequently, the floor settled, causing certain plumbing pipes under the concrete floor in the showers and toilets to break. This resulted in flooding and further settling. It was contended that no proper footing had been placed under the floor and that this, in turn, caused the floor to settle and the pipes to break.

Certainly, with all of these defects abounding, it was with a degree of trepidation that investigation on behalf of the architect into the background of these matters was commenced. The first order of business required the retention of an outside expert to furnish an opinion of liability. This proved to be an extensive task, but ultimately a written report was received, digesting the various claims and furnishing a detailed analysis of each.

Insofar as the architect's plans were concerned, the expert report contained the following commentary:

> The plans reflected standard architectural design relative to construction requirements employing the normal practice of using shop drawings prepared by the material suppliers. Traditionally, material suppliers (concrete reinforcing steel, structural steel, etc.) submit shop drawings to the general contractor showing complete details for construction. Upon review and approval by the general contractor these shop drawings are then submitted to the architect for his approval. The architect's approval of these shop drawings indicates that he has checked all solutions as well as details for conformance to the design requirement. Thus, until the architect approves *all* shop drawings prepared by each material supplier, his design is not completed. Construction generally does not begin until the contractor has received structural type shop drawings approved by the architect.

The expert concluded that the structural plans of the architect were not at fault; rather, he determined that the problems encountered were caused by changes made by the owner and/or contractor in the field during construction, which varied from the plans prepared by the architect.

Armed with this information, counsel for the architect went forward to defend against this action by the owner. During the remaining pretrial period, and during settlement conferences with the judge who was to try this case, the owner was made to realize that his claim for damages against the architect, though initially capable of supporting a substantial recovery, could be successfully defended against. Before the trial could be held, it was learned that the owner had sold the apartment complex and, although he contended that he had not made a profit, agreed to settle this action for the relatively nominal sum of $14,000.

TECHNICAL ANALYSIS

The first issue in this case is the deficiency in the size of the Lally columns. The contract plans indicated a diameter (3½ in), but no weight. The contractor (who was, in fact, also the owner) did not ask for a clarification, but installed the lightest weight section. Analysis indicated that, if everything else had been perfect, the lightweight section could have held the load. This is expecting too much. As noted in the Narrative, at the building inspector's behest, heavier sections were actually installed. Still, one or two columns bowed and had to be plated. The extent to which shoddy workmanship may have contributed to the problem was never resolved, but if the apparent deficiency was detected by the building inspector merely on the basis of a visual inspection then it must have been a gross deficiency indeed. Curiously, the architect insisted that it was the responsibility of the contractor to indicate the weight of column required.

The cause of the second issue, the sagging of the ceiling over the lobby, was never identified. It may have been due to failure to design for the weight of the nonbearing partitions which were located on all floors above the lobby, but which were omitted in the lobby space (see Figure 15-1).

The cause of the settling of the clubhouse floor was variously attributed to the lack of compaction of the soil under the slab and to the failure to provide footings under the interior masonry partitions. It was claimed that the 4-in slab was inadequate to function as a footing. The plan of the clubhouse is shown in Figure 15-2. The matter was never resolved.

The cause of the lean of the foundation walls can be seen by reference to Figure 15-1. It appears that the walls may have been backfilled before the floor system was in place to provide necessary anchoring and bracing.

LEGAL ANALYSIS

Before a complete legal analysis was possible in this case, it was necessary to understand the role played by the parties and the services each performed. On this project, the owner acted as his own contractor and was

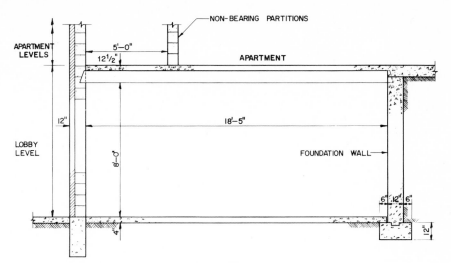

FIG. 15-1

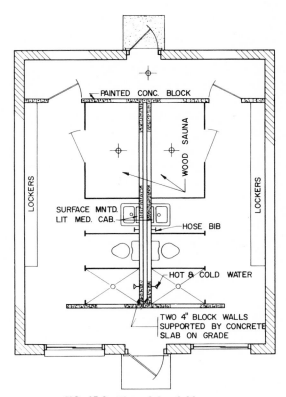

FIG. 15-2 Plan of the clubhouse.

instrumental in making decisions throughout the design and construction process. It is not uncommon for an architect to prepare all plans and specifications on a residential complex of this type. Usually, structural design of this nature is not complex and can be handled efficiently by the architect's own offices.

The architect's contract, however, did not provide for inspection services during construction. In effect, a design professional without a construction-phase inspection obligation has no assurance that the final product will conform to the design approved by the owner. There are architects and engineers who refuse to execute a contract for services which does not include the right to ensure compliance with their plans and specifications by the contractor through periodic or resident inspection. They argue that, unless they can control the final work product, they are at the mercy of owners and contractors who take liberties with the design and often transform it into something completely at variance with the design concept initially conceived and ultimately approved by the owner. However, it should be noted that not all design professionals have the luxury or the economic freedom of choice to insist upon this right and owners often choose not to assume the additional cost of having an architect or engineer on site during the construction phase.

On this project, cost-conscious decision making by the owner resulted in numerous changes by the contractor during construction. However, at the outset of the case, there was a definite belief on the part of the architect that he faced exposure because his original plans and specifications did not designate detailed specifications which had, in fact, resulted in failure. The most glaring of these concerns involved an alleged failure to have designated the particular type of Lally column to be used. Although this procedure was ultimately determined to be standard practice within the construction field, the architect did leave himself open to criticism by failing to make a designation of this specific type of column in his plans.

Many architects and engineers will conclude from this reading that detail of this kind cannot be incorporated into every project on a cost-effective basis. Perhaps this is so. However, it should be noted that when a claim arises, every aspect of the design process can be subject to question, and if a particular practice is deemed not to conform with the custom and standard of the industry as practiced in the locale in question, a successful claim for damages may be maintained, notwithstanding the fact that the architect or engineer may customarily do the same thing on numerous other projects.

Comments

It is common practice in residential work for the architect to do the structural design as well. The laws of some states permit this. Indeed, the definition of architect and engineer as set forth in the licensing laws may be substantially identical.

A second practice brought out by this case is a tendency by a builder to ignore the architect or engineer once construction has started and to take care of problems by himself, generally opting for the minimum-cost solution. Few sets of plans are perfect. Wise contractors (and owners) recognize that it is in their own interest to keep the designer involved— because it is the designer's best interest to keep the "job" out of trouble. A contractor may be an engineer by training, or employ capable engineers who well understand the design. Others show what, at times, seems to the designer to be a gross lack of understanding of what makes the structure function. Nevertheless, as the one primarily responsible for the design, a nondelegable duty evidenced by the seal on the plans, a design professional would be ill-advised to permit changes to his or her plans without a complete review and understanding of the change requested and its potential effect on the total design.

case 16

TYPE OF FACILITY   Hospital

TYPE OF PROBLEM   Settlement of foundation

Significant Factors   Failure to recognize the significance of the presence of 15 ft of organic soil (peat and muck)

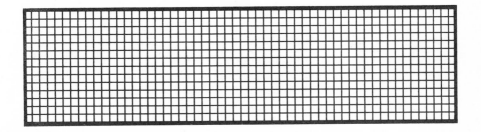

NARRATIVE

This case involves foundation settlement of the corner of an addition to an existing hospital facility. The cause of the settlement became quite clear upon investigation. The architect, who had his own structural engineers on staff, was retained by the hospital to prepare all plans and specifications, including foundation design. Prior to completion of the design, the architect was furnished with a soils report prepared by a soils engineering firm. The soils report contained recommendations about the type of foundation to be used in certain portions of the hospital. Specifically, the soils report noted that a certain area of the building was underlain by compressible soil material and that either (1) this material had to be removed or (2) that portion of the building had to be supported on pilings.

The ultimate design produced by the office of the architect did not provide for either treatment, and settlement of up to 8 in occurred in this portion of the building, as predicted.

The hospital embarked on a program to determine the cost of the repair that would be necessary to cure the settlement problem. A cost of approximately $400,000 was accepted as realistic. Repairs were effected under the direction of the architect, with the construction manager for the project acting as general contractor for the repair work.

In an effort to resolve the claim without recourse to litigation, an agreement was entered into whereby a "fact-finding forum" was set up among the involved parties to review questions of liability and damage. The concept of such a forum called for each of the parties to conduct its own fact-finding and subsequently meet with the others in an effort to reach an agreement about the facts and causation of the settlement. In this fashion, they hoped, settlement negotiations and a reasonable apportionment of liability would be agreed upon without recourse to lengthy court proceedings.

A fact-finding forum ultimately was arranged with representatives of the architect, the construction manager, the soils engineers, the concrete and steel subcontractor, the excavating contractor, and the hospital, and with attorneys for the various parties. Representatives of the hospital took the position that either the architect did not understand the significance of the organic soil in the area or the designer overlooked or ignored the condition. It was further stated that the soils report recommended that pilings be placed in the southeast area of the building (the area which had settled) and that, although the report's recommendation for piling along the northeasterly column lines of the building was followed by the designer, the recommendation for pilings in the southeast corner was not in fact carried out. The hospital argued that it was either the intention of the designer to remove the organic soil and replace it with compacted fill, thus obviating the necessity for piles, or merely that the piles had been negligently omitted from the design.

According to the hospital, although the soils engineers were called upon to test the soil compaction beneath the footings placed in the area that should have had pilings, they neither once raised an objection nor questioned the lack of pilings, despite their obligation to do so. Criticism also was leveled at the soils engineers' failure to spot the allegedly improperly compacted fill beneath the footings.

The hospital criticized the construction manager for failure to note the settlement during construction. It had been pointed out to him that the southeast column on the corner of the hospital was short by several inches, requiring additional material to be added. This condition, though known to the construction manager, was not communicated to the architect.

The various subcontractors were criticized for not having called attention to the short columns and settled foundations (which had existed for nearly a year prior to identification of the problem).

Representatives of the soil engineers, construction manager, and subcontractors all denied any liability. Each of these parties asserted that the architect, by his failure to specify pilings beneath the corner of the hospital, was responsible for all resulting damages.

As a result of this fact-finding forum, private negotiations were undertaken between the architect and the representatives of the hospital. The purpose was to acknowledge that the architect would have to bear the overwhelming responsibility for any damage and to determine whether additional contributions could be secured from the other parties to the project.

Ultimately, a settlement between the architect and the hospital was reached for the sum of $408,000. Separate settlement agreements were entered into by the hospital and the other parties for approximately $20,000.

TECHNICAL ANALYSIS

The presence of up to 15 ft of organic soil (peat and muck, with moisture contents of up to 200 to 300 percent) under one corner of the building was clearly noted in the soils report prepared for this project. The need for piling also was noted, but not so clearly; the designer did not provide the piling. As a result, the weight of the building and of about 16 ft of fill placed to raise the level of the site caused excessive settlement of the corner of the building, structural distress, and the peripheral problems noted in the Narrative. It is not clear why the designer did not provide pile support (he used spread footings founded in the fill). A failure to have understood—perhaps even to have read—the soils report may be the reason.

Remedial measures included underpinning the affected columns using augered, cast-in-place concrete piles. Selection of this type of pile was based on the desire to minimize vibration influence on the rest of the building. In addition, the sewer line servicing the building was replaced (on a pile-supported cradle) and the section of roadway serving the building entrance was also put on a supporting structure. The magnitude of the deficiency of support can be appreciated from the fact that the remedial supporting piles were designed for 45 tons of downdrag per pile. Some peripheral matters are of interest in this case:

1. The boring logs were not reproduced on the plans. It is usually desirable to do so, not only for information to prospective bidders, but because the plans come under many eyes in the process of implementation and many an engineer has benefited from an alert review by a contractor's employee or by the building inspector.

2. The settlement problem was not due solely to consolidation of the organic soil. Poor compaction of the fill material under the footings added about 2 in, or more.

3. As the building was being erected, the concrete contractor kept noting that the columns he previously had constructed were "low." No one attached any significance to this occurrence.

Finally, an interesting technique was used to evaluate the "locked-in" stresses in the structure of the building. These stresses were due to the settlements which had occurred. If the locked-in stresses were high, an argument was made that the building would have to be releveled (by jacking) to relieve the stresses before the live load stresses were added. How this could have been done was not discussed. The technique used to evaluate the locked-in stresses was to expose some of the reinforcing bars, attach strain gauges to the exposed bars, cut the bars, and record the relief of stress.

LEGAL ANALYSIS

Clearly, the most notable aspect of the case from a legal standpoint concerns the novel approach of utilizing a fact-finding forum as the means to resolve this dispute. The use of such a procedure will not lend itself to most construction problems. Generally, the concept will be most acceptable when there is no dispute over the existence of a defect, when there is an amicable working relationship among the parties, and when there is a mutual desire to seek an early resolution of the dispute without resort to the cost of litigation or arbitration.

Here, the hospital representatives were appreciative of the time and effort which had been expended by the architect in an effort to resolve the problem. The hospital also recognized that to attempt to reach a settlement by amicable means would be advantageous, since the cost of the remedial repair work involved the additional cost of debt service. As an inducement to the parties to participate in the fact-finding concept, the hospital stated that it would not attempt to seek consequential damages for any delays which were occasioned by the negotiated settlement.

The arrangement that was agreed to in this case called for each of the parties to prepare its own factual presentation and analysis of the causation of the problem. Following the presentation of these findings, settlement discussions could then be held as a means of bringing about a negotiated settlement with contributions from all responsible parties. In the event that a negotiated settlement could not be reached, a shortened arbitration proceeding was considered with a decision to be rendered by arbitrators. All statements or admissions made by any of the parties in the fact-finding forum were to be without prejudice, and none of the parties would use any admissions or statements against any of the others in any subsequent arbitration or litigation proceedings.

The concept of the fact-finding forum can therefore facilitate settlement of certain types of construction problems and should be considered as one of many ways in which to reach a negotiated settlement and avoid the customary adversary proceedings of arbitration or litigation.

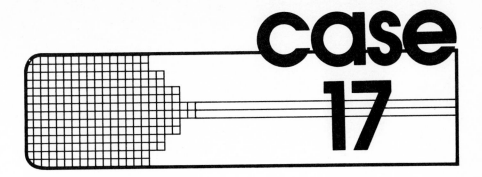

case 17

TYPE OF FACILITY   Warehouse

TYPE OF PROBLEM   Settlement of floor slab (slab on grade)

Significant Factors

A. Failure to compact fill under slab
B. Decomposition of organic material in the fill

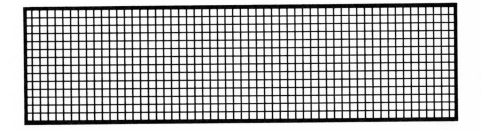

NARRATIVE

The structure involved in this case is a single-story warehouse. The warehouse facility was part of a construction project consisting of offices and a distribution center. Construction costs were in excess of $650,000.

Following completion of the building, the owner reported that the ground-floor slab had begun to sink. Further inspection confirmed that the concrete floor had, indeed, sunk, resulting in separation between the walls and floors. The net result was that partitions were literally hanging from the roof, rather than being supported on the floor.

An action was commenced by the owner against the architect, who in turn brought third-party actions against the general contractor, the pile driving subcontractor, and an engineer who had been retained by the general contractor to perform certain tests, including concrete strength tests on cylinders prepared during construction. In addition to out-of-pocket losses for repair of the building, the owner demanded $100,000 in damages for diminution in the value of the structure.

During the course of the discovery phase of the litigation, the architect attempted to point a finger at his structural engineer, stating that he relied upon the latter exclusively in all matters associated with the foundation. The architect further testified that the structural engineer was of the opinion that the settlement was caused by the poor quality of the concrete used in the floor slabs and by inadequate compaction of the soil and crushed rock under the floor slab (both were the responsibility of the general contractor).

For his part, the structural engineer contended that the problems were associated with the design of the architect and agreed with the architect that the problem also rested with the general contractor. No direct evidence was ever introduced to implicate the engineer retained on the part of the contractor.

A motion for summary judgment was filed on behalf of the engineer for the contractor, asserting that his services were limited to (1) observation and recording of the pile driving activities and reporting on this to the general contractor and (2) running compression tests on concrete cylinders. It was pointed out that the problem involved in this matter related not to the structural members of the building (neither the columns nor the piles on which the columns were placed were deemed defective) but to the nonstructural floor slabs which appeared to have insufficient support.

The general contractor argued against the motion for summary judgment indicating that witnesses would be called at the trial to testify that during construction there was a record of a rise in the groundwater level of as much as 16 to 18 ft, that this displacement of the water table could have caused the observed settlements, and that the rise of groundwater level could have resulted from the pile driving operation. This information was taken from a report which counsel for the contractor had not placed in the court record, preferring to use it at time of trial.

After due consideration, the court granted the motion for summary judgment, dismissing the engineer for the contractor. Shortly thereafter, with trial imminent, a settlement was negotiated by the owner with the general contractor and the other parties to this action.

TECHNICAL ANALYSIS

The site of the building was underlain by a 6 to 10 ft thickness of "trash" fill (ash, glass, sand, tin cans, paper, rags, wood, scraps of plastic). The soils report which was developed "after the fact" noted a strong discharge of marsh gas from the fill and a loss of weight on ignition of samples of the trash fill varying from 5 to 40 percent, with an average of 11 percent. The organic material simply was rotting away (turning into gas), leaving voids in the ground, and causing the floor slab which was supported on the ground to settle as the fill collapsed into the voids. About 2 to 3 in of settlement had occurred by the time it was decided to take corrective action. If such action had not been taken, approximately 18 in of settlement would have eventually occurred.

It was clear that proper design would have involved either structural support for the floor slab or the removal of the trash fill and replacement with sand and gravel.

The problem was compounded by the fact that the under-floor fill, according to density tests performed during development of the defense of this case, had received *no* compaction other than from spreading and grading.

Initial attempts at remedial measures consisted of pumping 200 tons of cement grout under the floor slab in an unsuccessful effort to compensate for and stop the settlement (a failure which should have been clearly anticipated). Following this failure, a proper "fix" was instituted. A system of pile supports was installed for the floor slab, as shown in Figure 17-1, and a new slab was poured on top of the settled slab. The doors and walls were straightened.

Comment

A foundation engineer, and civil engineers in general, upon reading this case (and Case 16) will be aghast that the events described herein could have been permitted to happen. The development of the problems described should have been obvious. But, this case is the third instance of its type that the author has encountered—same type of building, same type of fill, and same result. The reader should note: Organic soils are

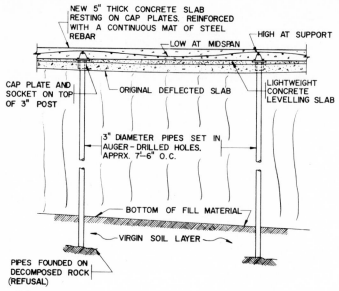

FIG. 17-1 A system of pile supports installed for the floor slab.

potential sources of problems. Fills are potential sources of problems. Fills with organic inclusions (or trash) are double sources of problems.

LEGAL ANALYSIS

As many design professionals will know from their own experience, there has been a growing trend for plaintiffs in construction litigation to name each and every party associated with the project as a defendant when litigation ensues. True, there are instances when this "shotgun" approach represents a rather callous and indifferent attitude toward clearly innocent parties who bear no relationship to the basis for the litigation. Perhaps some comment is appropriate as to the bases upon which attorneys will name parties who, in the first instance, may seem removed and apart from the subject of the litigation.

When a design or construction problem appears, counsel will generally take a broad view of the range of possibilities, considering any and all possible causes. Failure to consider all possibilities can often result in pursuing parties whose responsibility is minimal, or even nil, while omitting parties with real liability. In many instances, local statutes of limitation will require that litigation be brought against architects and engineers within a specified time period or constitute a waiver of any right to proceed against them. It is for this reason that attorneys will generally join more, rather than less, of the parties potentially responsible for the problem at hand.

Moreover, it is often difficult at the onset of litigation to determine the precise cause of a particular problem. What appears to be a clear-cut analysis at an early stage of investigation may result in new information that leads to other, more direct, causative factors. As the cases in this book have shown, expert analysis often leads to new roads of inquiry and the introduction of new parties to a problem who had previously not been considered as involved. The dilemma facing an attorney representing an owner is to pursue, with expedition, the responsible parties and recover any damages sustained by the owner. There is often a fine line drawn between pursuing those who are deemed "responsible" in the eyes of the owner and pursuing defendants who, from their vantage point, have no reasonable relationship to the problem at hand.

In the case discussed above, certainly it was the position of the engineer for the contractor that there was no liability on his part in light of the limited scope of his duties on the project. Nevertheless, according to the attorney for the owner, there was some question as to the relationship of the engineer to the project as a whole and the structural components which had failed. Initial discovery did enable counsel for the engineer to narrow the issues, limit the inquiry to those specific jobs performed by the engineer, and determine that no testimony was to be put forth which would implicate the engineer as a cause of the problem. For this reason, the summary judgment motion was successful.

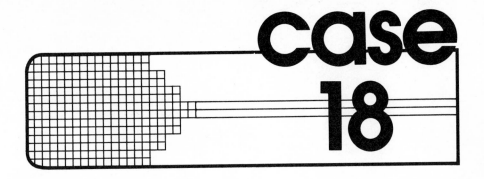

case 18

Grandstand

Collapse of roof

Significant Factors

Buckling of web of rigid frame knee due to inadequate stiffeners

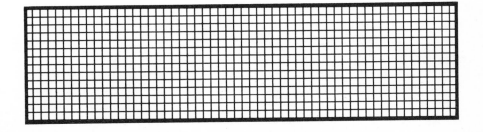

NARRATIVE

Though engineering is often referred to as a science, it is the subjective beliefs of the engineer which often have a dramatic impact upon the ultimate design product. These subjective impressions can come from a number of sources and cast an imprint upon the final design by permitting or persuading the engineer in subtle ways. A case in point involves the design of the grandstand portion of a sports facility. The architect entered into an oral agreement with the structural engineer to prepare the structural plans, but to assume no on-site obligations during construction.

Prior to preparation of the structural plans, the owner showed the structural engineer plans for a similar facility which had been constructed in another part of the state. The structural engineer was later to acknowledge that he was impressed enough by the design of the earlier facility to incorporate a similar concept for a cantilevered roof to be built at the new project. Although this project was to be much larger in scope, and would include additional stress on the roof from a press box which was to rest on it, he nevertheless chose to follow the earlier design, essentially without change. The structural engineer later admitted that he had not done enough research on the design to evaluate the additional stresses that would result.

The cantilevered roof, as designed, was to extend over a grandstand. The design was anticipated to be adequate for a 20-lb/ft² live load, i.e., adequate for any possible use intended by the owners. During construction, with all the framing erected, the metal deck installed, the roofing and insulation in place, i.e., with most of the dead load acting, but with no substantial live load, the roof collapsed at the supporting columns. The collapse developed gradually, over a period of several minutes. No personal injuries were sustained.

Immediately following the collapse, an investigation was undertaken on behalf of the structural engineer. This included the immediate retention of outside consulting structural engineers who visited the site and began a review of the design and construction of the roof. It was the conclusion of these outside experts, as well as experts retained by the owner, that no construction defect had caused the collapse. To the contrary, it was determined that the proximate cause of the collapse was an inadequacy of the design.

The design error was determined to have resulted from the structural engineer's failure to incorporate adequate web stiffeners into the roof girders at their junction with the columns (the "knee"). A web stiffener is a small structural member added to a slender beam or column to prevent buckling. Here, the stiffener should have been added at the center of the web, extending the complete width of the beam supporting the roof structure. The web at the joints between the columns and the girders buckled, collapsing the column section and causing it to warp and twist.

Further investigation disclosed that the girders had been properly constructed by the contractor and that fabrication of the columns, beams, and girders complied with AISC specifications.

Once it was determined that the fault rested entirely with the structural engineer, efforts were undertaken to avoid delay damages and to mitigate further repair costs by moving quickly to ensure that all remedial work was promptly undertaken. Fortunately, there was a period of some months before the facility was to be opened and this permitted removal of the collapsed deck, fabrication of new girders, and reconstruction on an accelerated time schedule. As a result, and in coordination with the outside experts who approved the remedial design, the facility opened on time and extensive delay damage claims were avoided.

Proven damages in excess of $350,000 for repair and replacement of the grandstand roof had been incurred by the owner. Settlement via negotiation was ultimately reached with payment on behalf of the structural engineer of $34,000.

TECHNICAL ANALYSIS

The structure under discussion is shown in Figure 18-1. The condition after the collapse is shown in Figure 18-2.

Web stresses in the knee at the cantilever of this frame are indicated in Figure 18-3a. Analysis (finite element analysis was used) indicated the extent of the tension and compression zones to be as shown in Figure 18-3b. In the compression zone, the web must act as a column, with a length of 72 in ("critical length for buckling") between the stiffening flanges. The compression stress on this section was calculated as 12.1

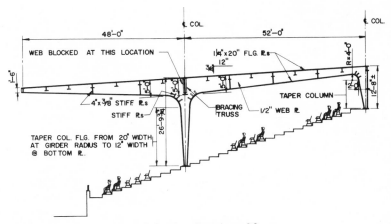

FIG. 18-1 Elevation view of frame.

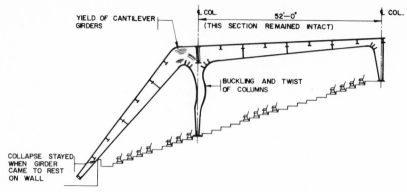

FIG. 18-2   Schematic of condition after collapse.

kips/in² under the loads actually in place at the time of collapse. The capacity (ultimate strength) was calculated (considering the web as a thin plate fixed at the ends by the flanges acting as stiffeners) as 8.5 kips/in². This capacity was increased by some undefinable amount by the bracing effect contributed by the bottom chord of the bracing truss and, to a degree, by the presence of the short radial stiffeners. Clearly, a very tenuous condition was present, as, indeed, the collapse evidenced.

Comments

The problem of web buckling in rigid frame knees is well known, and proper attention is necessary. Reference is made to the AISC publication *Design of Rigid Frame Knees.*

Corrective action consisted of removing the damaged sections of the roof and replacing them with new material. Additional web stiffeners were

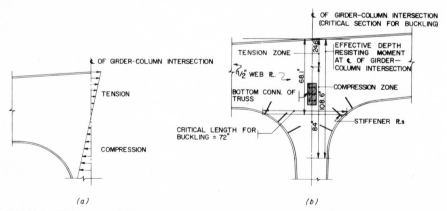

FIG. 18-3   Web stresses in frame. (*a*) Schematic. (*b*) Elevation view of girder-column intersection.

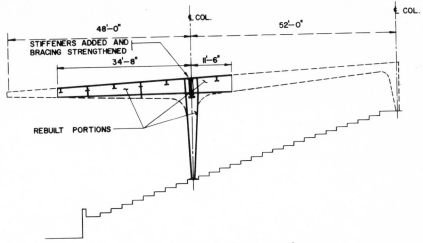

FIG. 18-4 Schematic of reconstruction.

provided at the girder knees, and the lateral bracing system for the joint which collapsed was strengthened. Perhaps most important, the cantilever overhang was reduced from 48'0" to 34'8" (see Figure 18-4).

The expert's review of the collapse properly went beyond investigation of the collapsed framing. He reviewed the adequacy of the framing below the grandstand as well. This review brought out another significant factor; again well known, but, as this case demonstrates, not always appreciated. That factor was the overstress condition caused by eccentricity in connections. The condition is shown in Figure 18-5b. Calculations showed bending stresses under full dead and live loads to be up to 94.2 kips/in² and shear

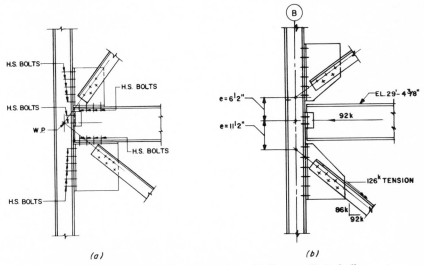

FIG. 18-5 Wind framing connections. (a) Proper. (b) As built.

stresses to be up to 40 kips/in² (far excessive for A36 steel) due to the built-in joint eccentricities.

LEGAL ANALYSIS

There is little in the way of legal analysis that can be imparted to a case of this type. However, a number of lessons can be learned from total failures which are clearly the result of defects in the design drawings.

The attorney representing an engineer encounters a shocked and defensive client who has just been informed that a structure he or she designed has collapsed. This is a very vulnerable time for the engineer, who must be approached with a certain sense of caution and consideration owing to the sensitivity of the situation. Quite often, the engineer has already reviewed the plans and has, in his or her own mind, come to a conclusion whether an error exists. Sometimes, feeling fully at fault, the designer is ready to admit an error before a detailed analysis of the problem can be completed.

It has been our experience that an attorney in this position should not rush to blindly accept the conclusions drawn by the design professional. Rather, the attorney should calmly and dispassionately approach the situation in a manner which will elicit the facts in as orderly and professional a manner as possible under the circumstances. This will include carefully reviewing all information provided by the designer, meeting with other involved personnel, early involvement of an outside consultant, and analysis of all necessary documentation and code requirements.

On more than one occasion, we have been confronted by design professionals who have readily admitted liability for a failure, although further analysis disclosed no liability in regard to the design documents. Though such situations are rare, it is incumbent upon the attorney to avoid any premature conclusion which can have an adverse effect both upon the designer and the designer's relationship with the owner and other parties.

In this case, the acknowledgment of liability was buttressed by outside experts who completed a prompt analysis of the failure. This enabled discussions with the owner and other parties to revolve around determining the most cost-effective method for removing the debris and repairing the damage. This helped to avoid delay damage costs which could have been demanded by the owner had the sports facility not opened on time. As we have noted earlier, it is as important for the attorney to effect a course of action designed to mitigate damages once liability is fixed as it is to determine the cause of the failure.

The "success" achieved in this case revolved around the efforts to pinpoint the cause of failure, effect early repair, minimize the damages incurred, and accomplish all of this in a manner that avoided protracted and costly litigation.

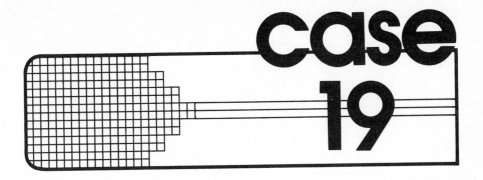

case 19

TYPE OF FACILITY

Single-family residence

TYPE OF PROBLEM

Settlement resulting in cracking of partitions, cracking of floor slab, and broken utility lines

Significant Factors

A. Permafrost
B. Failure to explore subsurface conditions (until after problem had developed)

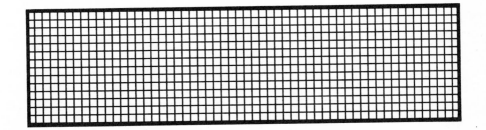

NARRATIVE

This case involves a problem peculiar to northern climates, particularly those areas of the country which face subfreezing weather for a substantial portion of the year. In these locales the problem of permafrost must be considered in planning a foundation design which will insure against structural damage from settlement movement after completion of the structure.

Permafrost is frozen earth material located below the surface. If thawed, such ground material moistens and softens, resulting in settlement of any structure supported by the formerly frozen material.

Shortly after the construction of a single-family residence, the home sustained settlement damage which, at the time, was attributed to failure of the water well. The pipes had cracked. Freezing and thawing of water from the broken pipe resulted in buckling of the slab on grade as well as erosion of gravel from underneath the footings of the house. The owner submitted a claim to his homeowners' insurance company, which paid $9000 to effect repairs. An inspection by the developers indicated no evidence of permafrost, it being the belief of the owner and the insurer that the broken pipe was the cause of damage.

Prior to purchase of the home, the owner had requested an inspection by an engineer to determine the existence of permafrost. There was some apprehension on the part of the owner since he had been aware that some of the lots had evidenced patches of permafrost. A brief inspection, taking approximately 1 hour, resulted in a letter from the engineer to the prospective owner stating that although scattered pockets of permafrost were known to exist in the area, none was believed to exist on the particular lot to be purchased.

Shortly after the purchase, the pipe break occurred and, as noted, was attributed to the well failure. It was only after a new well was constructed that evidence of permafrost was found. This subsequently resulted in an action against the engineer for damages to the property deemed to be in excess of $100,000. The suit also sought to recover damages from the prior owners for misrepresentation and breach of the warranty contained in the deed, on the grounds that the existence of permafrost should have been known to, or detected by, the prior owner and disclosed prior to sale.

The engineer stated that, in addition to his letter, he advised the owner that the only true way to determine whether permafrost existed on the property was to take borings. According to the engineer, the owner indicated that he did not wish to incur this expense. It should be noted that the engineer was not a soils engineer, but at no time disclosed this fact to the prospective owner.

The defense of contributory negligence was asserted by the engineer in his answer to the complaint. It also was asserted that this claim was covered by the plaintiff's homeowners' policy, which had earlier made payment for the pipe damage.

During the course of the litigation, the owner sold his house, incurring a loss estimated at $20,000. The issue in this case centered around whether the settlement of the house was caused by the permafrost or by the prior damage which had occurred beneath the home. An independent expert retained by the owner reported that permafrost existed at a depth of approximately 6 ft beneath the house and that the cost of correcting the structural damage to the residence could total $50,000. A second expert report commissioned by the owner indicated frozen ground at a depth of 18.5 to 19 ft. This report further indicated a belief that at the time the residence was built, permafrost was much closer to the ground surface. As heat from the home escaped, insulating the ground beneath from the winter cold, the permafrost retreated to a greater depth.

An expert retained on behalf of the engineer could not refute the findings of the opposing expert testimony, in effect leaving the owner to pursue his further claim for misrepresentation against the former owner.

TECHNICAL ANALYSIS

The technical problems in this case are basic. Borings, made *after* the event, showed nonplastic and organic SILTS to depth. The soil was frozen at a depth of about 19 ft. A layer of ice was encountered at a depth of 21 ft. Logs of two of the borings are presented in Figure 19-1. Progressive thawing of the frozen soil and consequent settlement of the building was clearly the problem. About 6 in of total settlement occurred, as shown in Figure 19-2. The solution was a freeze-back system to prevent additional thaw.

Of greater interest was the sequence of events which led the engineer into difficulty. At the time that the owner of the residence was considering its purchase, he called in an old friend, a principal of a consulting engineering firm, to give an opinion about the soundness of the construction. The friend sent one of his staff of engineers to inspect the building. The inspection consisted of a walk around the exterior; examination of the interior for visible defects, such as cracks in the partitions; opening and shutting the doors and windows to check for binding; and a check of the foundation walls and of the concrete floor slab (where exposed) for cracks. The inspection did not disclose anything unusual. The report of the engineer stated the following:

> I could find no visible evidence that the building is structurally unsound. The fact that scattered pockets of permafrost are known to exist in the area is cause for concern. However, I do not believe there are any on this particular lot. The growth of large diameter birch and spruce are good indications that the immediate area is not frozen. I also understand that the building

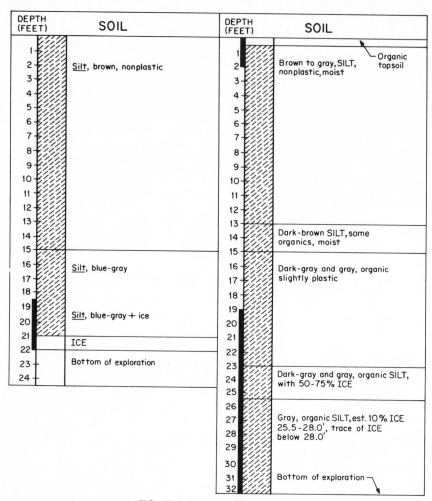

FIG. 19-1 Logs of two borings.

is some two or three years old, which should be old enough for any foundations problems to become evident.

He was wrong! Was it bad luck or bad judgment? One or two borings would have resolved the issue. The engineer's reluctance to impose the cost of such boring(s) on a small home owner is understandable—but in this case, and in other cases, it is necessary. At least one building code (New York City, 1965), based on a history of unfortunate experiences, mandates subsurface exploration for single-family, as well as multifamily, dwellings.

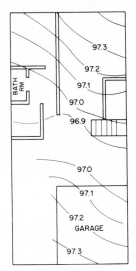

FIG. 19-2 Plan showing settlement contours.

LEGAL ANALYSIS

Pointing out the strengths and weaknesses of the engineer's case in this matter and arguing them to conclusion could be the subject of a moot court competition. First, there clearly was a strong belief that the owner was making a fraudulent claim, having previously submitted his damage claim for settlement to his homeowners' insurance company and having already received payment for it. Further, there was testimony from the prior owner that full disclosure of cracks and other problems with construction of the house had been communicated to the purchaser prior to sale.

However, there was an entirely different side to this argument: the fact that the engineer's cursory report following his inspection specifically stated that no permafrost existed on the premises. This was clearly refuted both by experts retained by the owner and by an expert retained in defense of the engineer. Moreover, since the current owner was not the original one who bought the house from a developer, the primary cause of the settlement appeared to be based upon the misrepresentation of the condition of the house at the time of sale.

Separate causes of action were pleaded against the former owner and the engineer, the latter for his negligent inspection and preparation of a report which had been relied upon by the owner with respect to the condition of the house. Consequently, counsel for the engineer was faced with the prospect of deciding whether to seek a negotiated settlement of this case or to proceed through trial, seeking a defense verdict on behalf of the engineer.

A further problem was the nature of the liability of the engineer, who contended that the purpose for which he was retained was to determine whether the owner's residence was structurally sound, not to advise about the existence of permafrost. It was his opinion that at the time of his inspection the building was structurally sound. However, the fact that his report contained remarks regarding the existence of permafrost undermined this defense, since the owner alleged that he relied upon the statement in the report as to the nonexistence of permafrost.

Finally, the issue was raised whether the further settlement of the house was additional damage traceable to, and resulting from, the broken pipe (which had caused the damage covered under the homeowners' policy), or whether this was a separate and unrelated incident. According to the experts, there were conflicting viewpoints. This issue, more than any other, was vigorously pressed on behalf of the engineer and may have been the primary reason for the owner's decision to accept a $30,000 negotiated settlement rather than proceed to trial.

From the vantage point of the reader, a case of this nature points up the quagmire design professionals can become immersed in when they gratuitously offer an opinion without taking all affirmative steps to ensure that their opinion is based upon facts. As is pointed out in the Technical Analysis section of this chapter, borings would have confirmed the existence of permafrost under the house. If the engineer chose not to take borings, which he knew were necessary if he were to give an assured opinion on the existence of permafrost, he should not have rendered an opinion in this area. Rather, he should have recommended in writing that these borings be taken and, if the owner chose not to accept this advice and the cost that went with it, should have placed a memorandum in his file reflecting the owner's reluctance.

Attorneys are frequently confronted with similar situations, and their best advice in those circumstances is that attention be given to keeping appropriate memorandums of decisions during the course of the project so that there is a record to review in the event of problems at a later stage. Of course, design professionals, like all other business people, are wont to keep voluminous records and memorandums. Accordingly, discretion is always advised, and perhaps a case such as this one best points out that good judgment may be a greater factor in preventing claims than many would wish to acknowledge.

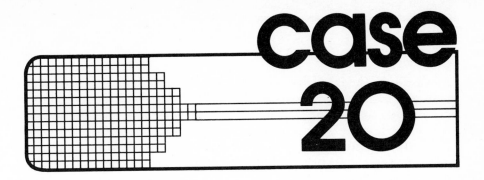

case 20

TYPE OF FACILITY
Sludge digester tanks in sewage treatment plant

TYPE OF PROBLEM
Cracking and spalling of corbel supports for floating roofs of tanks

Significant Factors

A. Design of corbels
B. Thermal effects added to stresses due to primary loads
C. Faulty construction procedures

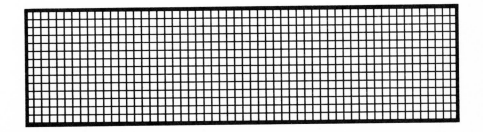

NARRATIVE

This case, as Case 4, involves a problem with the design of corbels. The structure is a set of digester tanks at a sewage treatment plant. The tanks, as is common practice, are provided with "floating roofs," which rise and lower as the level of filling in the tanks varies (see Figure 20-1). There is a lower stop to prevent excessive lowering of the roof. In this case the stop is a series of corbels around the interior face of the walls of the tank (Figure 20-2).

Cracking and spalling of the face and edge of the corbels were noted during erection of the floating roof—with only the roof trusses in place and none of the concrete decking or the roofing as yet erected, i.e., with no live load acting and under a total load about one-third that for which the corbels had been designed.

Initial investigation indicated that the corbels had strength well in excess of that required to support the full roof load and, of course, far more than what was required to support the partial loading under which cracking and spalling had developed. Further, two other plants were found to be in service with the same corbel design and these plants were found to have had no problems related to the corbels.

The municipality which owned the treatment plant considered various remedial repair schemes and subsequently completed the required repair work at a cost of $129,000. The different methods considered for the repair are shown in Figure 20-3*a* and *b*.

Following execution of the repairs, an action was commenced by the municipality against the engineer, who in turn brought a third-party action

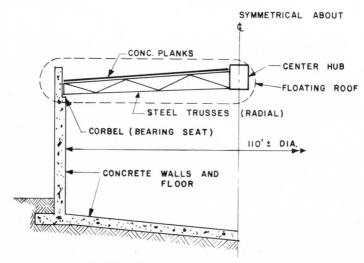

FiG. 20-1   Schematic section of digester tank.

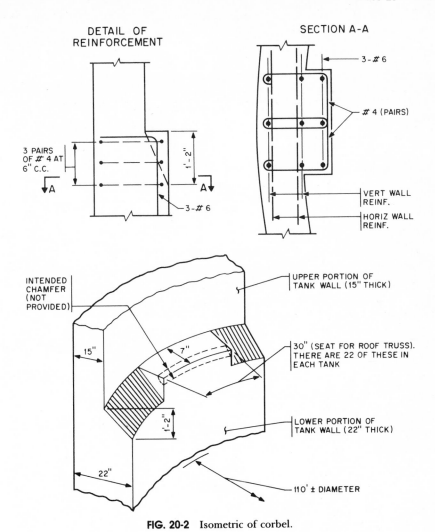

DETAIL OF
REINFORCEMENT

SECTION A-A

3 - #6

#4 (PAIRS)

3 PAIRS
OF #4 AT
6" C.C.

1'-2"

A

A

3 - #6

VERT WALL
REINF.

HORIZ WALL
REINF.

INTENDED
CHAMFER
(NOT
PROVIDED)

UPPER PORTION OF
TANK WALL (15" THICK)

15"

7"

30" (SEAT FOR ROOF TRUSS).
THERE ARE 22 OF THESE IN
EACH TANK

1'-2"

LOWER PORTION OF
TANK WALL (22" THICK)

22"

110' ± DIAMETER

FIG. 20-2   Isometric of corbel.

against his consultant, who had done the actual design of the corbels. This consultant turned out to be without professional liability insurance and otherwise insolvent and therefore primary liability was thrust upon the engineer for any design negligence.

During the course of the litigation, experts were produced who testified that the corbels failed not primarily as a result of vertical loads (i.e., the weight of the covers) but as the result of horizontal loads resulting from variation in the size of the covers due to thermal changes. Other, contrary, expert testimony was secured to the effect that for both vertical load and thermal effects, the design of the corbels was perfectly adequate and that

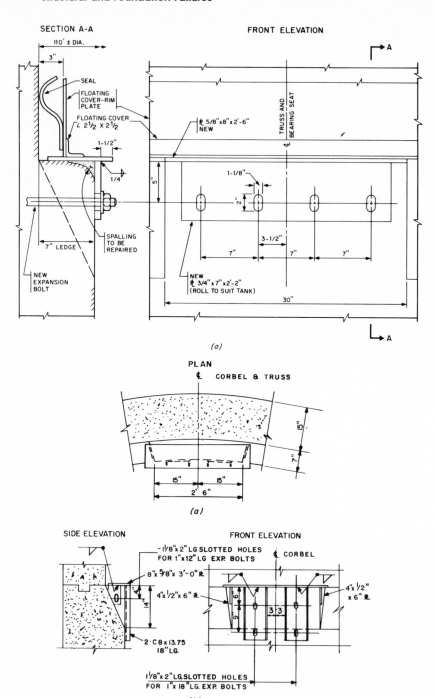

FIG. 20-3 Proposed method of repair. (*a*) Alternative 1. (*b*) Alternative 2.

the cause of the failure was related to the construction sequence and was the fault of the contractor.

It was interesting that the consultant who actually designed the corbels testified that the engineer had provided him with information which led him to design the corbels for the weight of a cover which was only one-fourth of the actual weight. However, he also testified at his deposition that the design of the corbels was adequate even with the weight of the cover at four times the value for which he had designed them, i.e., he had overdesigned for vertical-load by a factor of at least four.

Following the completion of pretrial activity, discussions were held with the municipality in connection with the possibility of settlement. The municipality, perhaps in recognition of the totally conflicting expert opinions which had been elicited, acknowledged that proving design inadequacy was subject to establishing greater credibility for its experts than would be established for the experts appearing on behalf of the engineer. The municipality also recognized that it would be required to show why the corbels at this plant failed when a similar design at two other plants in the vicinity had not. Accordingly, prior to trial, the municipality agreed to reduce its demand and accepted a settlement in the sum of $40,000. Because of the insolvency of the consultant, the engineer contributed the full amount, agreeing to accept a nominal contribution from the consultant in return for a full release.

TECHNICAL ANALYSIS

The overall structure is shown in Figure 20-1. Details of the corbel (bearing seat) are shown in Figure 20-2. An enlargement showing the position of the reinforcement in the bearing seat is found in Figure 20-4. Typical views of the spalling which occurred are shown in Figure 20-5.

The proximate cause of the failure of the corbels was never identified to the agreement of all parties. Accordingly, the following facts are presented without a conclusion.

1. At the time of failure, the vertical load acting on the bearing seat (corbels) was about one-fourth of the design load.

2. Analysis indicated that even without reinforcement the bearing capacity and shear resistance of the corbels was far in excess of the stresses imposed under the full design load. The vertical load alone could not have caused the failure.

3. From Figures 20-4 and 20-5, note that, in general, the depth of spalling is so shallow that the reinforcing bars intended to prevent its occurrence are not engaged (do not cross the crack). They were not effective in their function.

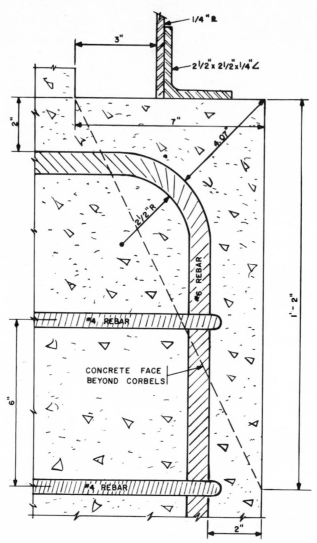

FIG. 20-4 Enlargement showing position of reinforcement in bearing seat.

With these facts in mind, the reader may consider the differing arguments[1] about the proximate causes which were presented in this case. These were as follows:

1. To guarantee proper fit of the assembled roof within the concrete tank, the specifications called for preassembly in the shop. The roof would then be disassembled ("knocked down") for shipment to the site. The

[1] Various "red herrings" which were raised are ignored.

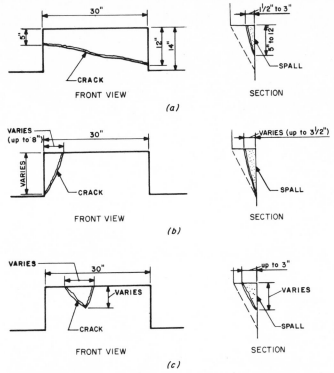

FIG. 20-5 (*a*) Overall spall of face; (*b*) Spall of corner of bearing seat; (*c*) Spall of center of bearing seat.

reason for this requirement was that the roof was intended to have a uniform, 3-in concentric clearance from the concrete walls (see Figure 20-3*a*, Section A-A), and so to bear near the middle of the bearing seat (corbel). The actual gap varied from 0 to 6 in. Where a larger gap existed, the weight of the roof was supported nearer to the edge of the bearing area and nearer to the unsupported and unreinforced edge of the corbel, beyond the concrete retained by the embedded reinforcing bars. It is difficult to credit this fact alone with causing the spalling, because drift of the roof from the concentric position during the rise and fall of the roof incident to operation of the digesters would have been expected. Some other, additional, cause had to be operative.

2. One such possible additional cause was inaccurate leveling of the bearing surfaces of the corbels. This condition was observed, with the bearing of the trusses concentrated near the edge of the bearing seats. Concentration of the bearing stress (and internal stress in the corbels) would result. The tendency to spall would be increased (see Figure 20-6*a*). A concentration factor of 20 percent was estimated by one investigator.

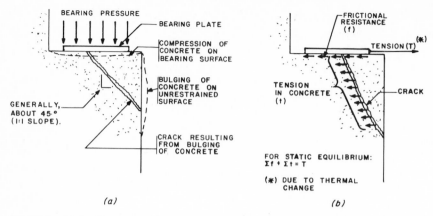

FIG. 20-6   (*a*) Spalling caused by vertical load alone. (*b*) Effect of induced lateral load.

3. The eccentricity of the point of bearing was exacerbated by the fact that the concrete walls of the tanks had been built "out-of-round" by ±1¼ in. Also, the edges of the bearing area had not been chamfered (see Figure 20-2).

4. A drag force (friction) due to thermal expansion and contraction of the roof was postulated and would have induced tension in the corbels which could have induced the observed spalling. The effect is illustrated in Figure 20-6*b*. The cracking shown in Figure 20-6*a* is due to vertical load alone and results from dilatance of the unsupported face of the concrete (so-called Poisson ratio effect). Theoretically, such cracks would be inclined about 1 to 1, as shown. The cracking shown in Figure 20-6*b* is more likely to be steeply inclined, as shown. The steep inclination of the actual cracks, as shown in Figure 20-5, was interpreted to be indicative of the effect of lateral forces.

5. With these eccentricities and the drag force considered, one investigator calculated the shear stress in the concrete near the edge of the corbels at about 75 lb/in². Allowable value for the 4000-lb/in² concrete used in the work is 70 lb/in². Ostensibly, an overstress existed. However, the design value of 70 lb/in² has a built-in safety factor of about 2½. Accordingly, these factors together could not have produced failure. Some other, additive, causes (faulty construction) had to be operative, such as those described in Paragraphs 6 and 7. Another investigator, however, calculated that the resistance of the concrete was grossly inadequate if the thermal drag was considered. His computation was based on the equations in the Prestressed Concrete Institute (PCI) design handbook *Precast and Prestressed Concrete,* which gives the formulas indicated in Figure 20-7. These conflicting conclusions led to the conflicting expert testimony noted in the Narrative section.

EQUATION FOR CALCULATING
CAPACITY OF CORBELS[2]

For uniform bearing on plain concrete, the ultimate bearing strength may be limited by

$$F_{bu} = C_r \phi 70 \sqrt{f'_c} (s/w)^{1/3}$$

where $\phi = 0.70$

$\quad s =$ distance from free edge to center of bearing (in)

$\quad w =$ width of bearing, perpendicular to free edge (in)

$$C_r = \frac{sw}{200} \frac{T_u}{V_u}$$

$T_u =$ ultimate axial tension

$V_u =$ ultimate shear force at section

$C_r = 1.0$ when $T_u = 0$

$f'_c =$ ultimate compressive strength of concrete

FIG. 20-7

6. Damage to the edges of the corbels had occurred when the forms were removed. The damages had been repaired. However, the observed failures were not limited to corbels which had been damaged.

7. A considerable amount of field welding of the trusses was required after they were erected. Thermal input to the steel, and subsequent shrinkage, had to result. The induced lateral force would have added to the drag due to downward variation in air temperature. There would be an upper limit value to the sum, however—the force at which slip occurs.

LEGAL ANALYSIS

At the outset of this litigation, various defenses were raised on behalf of the engineer. It was asserted that preliminary plans had been submitted by the engineer to the municipality, showing a continuous-ledge method of construction for support of the sludge digester tank covers. This design was rejected by the municipality in favor of a corbel design. Accordingly the engineer argued that the owner was estopped from asserting any alleged defect in the design of the corbels, since it had insisted upon and approved the design in question.

Further, it was alleged that the spalling which occurred was the result

[2] From: Precast Concrete Institute Design Handbook, *Precast and Prestressed Concrete*, p. 6–6.

of negligent construction on the part of the contractor, who was not a party to this action. Finally, it was asserted that the consultant, who had performed the actual design, was primarily responsible for any negligent design and that this consultant owed a duty to indemnify the engineer for any loss that might be suffered by the latter.

The insolvent position of the consultant, as well as his contention that he was provided with erroneous data by the engineer prior to completing his design, did raise a substantial question of fact which, if litigated, would only have served as additional support for the claim of the municipality. On this point the engineer's position was that this fact had nothing to do with the failure of the corbels, since the total weight of the roof at the time of the failure was approximately one-fourth of the total weight. The consultant was prepared to testify that the corbel design was adequate, even taking into consideration the correct weight of the roof.

Careful analysis of this claim resulted in an opinion that the municipality's chances of prevailing were somewhat less than 50 percent. Coupled with the insolvency of the consultant, the counsel for the engineer was able to project that a favorable settlement would be in the range of 40 percent of the total damages, i.e., 40 percent of $129,000, or the sum of $51,600. The ultimate settlement at $40,000, therefore, was considered highly favorable to the interests of the engineer, avoiding substantial trial and expert costs which would have been required in the absence of settlement.

Prior to finalizing the settlement with the municipality, the engineer reached an agreement with the consultant whereby the consultant obligated himself to pay to the engineer the sum of $5000 towards the settlement. Owing to the consultant's insolvency, the engineer agreed to accept payment of this sum over a period of time.

Taking all the exigencies of this case into consideration, the settlement that was effected was a reasonable one. Knowing that an owner will be able to put an expert on the witness stand at the trial who will testify to design error immediately puts into issue a question of credibility of the experts on both sides. No attorney can predict how a jury will react when it hears disparate opinions about the causation of a failure. Some attorneys will argue that a jury is generally sympathetic to an owner who has incurred a failure and the attendant cost of repairs. The question then is raised as to how a jury could deny a recovery, in whole or in part, to an aggrieved party in such a situation.

On the other hand, where an aggressive defense posture has been assumed by the design engineer, and where it can be shown that liability for the failure may be placed upon the owner or his agents, i.e., the contractor or construction manager, counsel for the owner should not cavalierly overlook the possibility of a defense verdict.

For these reasons, a careful analysis of the strengths and weaknesses of one's case often will lead to a reasoned analysis for purposes of settlement. Overlooking each of the various defenses of an architect or engineer can be fatal to such an analysis on the part of an owner. Certainly, it can often prevent fruitful settlement negotiations which, inevitably, are in the best interests of all parties.

Conversely, total reliance upon favorable expert reports to defend a claim of design negligence against an architect or an engineer is naive and reckless. Every attempt at apportioning the potential risks of litigation and proceeding to trial should therefore be carefully considered so that all possibilities may be discussed with the client.

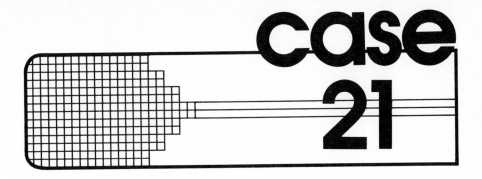

case 21

TYPE OF FACILITY   Stadium (stands)

TYPE OF PROBLEM   Structural inadequacies in columns, lateral bracing, and connections

Significant Factors
_____

A. Poor detailing of steel framing

B. Plans and computations not checked

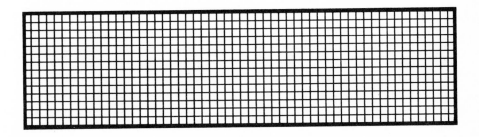

NARRATIVE

At one time or another, every architect or engineer has encountered a project which, from beginning to end, was rife with problems, complexities, and turmoil on the part of all concerned. Such situations usually manifest themselves at an early stage of the project and portend difficulties that threaten to engulf the unwary. Unless the design professional is prepared to gain an early hold on these problems and assume control of events, he or she can anticipate that matters will become more complex as the project proceeds. A general rule that can be established on construction projects, as in all business operations, is that problems will not disappear of their own volition and that if not tended to, they will reappear at a later date in a magnified proportion.

Such was the case with a municipality whose goal was to enlarge and enhance its antiquated football stadium to equal that of other large stadiums capable of supporting a professional football team. At the outset, a turnkey developer was brought in and given a contract to expand the stadium within a budget of $2.9 million. The developer emphasized its experience in stadium design and its ability to bring a project to completion within a modest cost of construction. It later would be learned that this developer had never designed any prior stadiums of the type or size envisioned, and that the only related similar type of projects which had been designed and constructed by this developer had involved single-tier grandstands rather than the multideck stadium contemplated.

The developer entered into a contract with an engineering joint venture which was to prepare the plans and specifications and oversee construction. In turn, the engineering joint venture (hereafter called Jamison) retained an engineering firm, to prepare all design work for the structural system for review and approval by the turnkey developer.

Under a tight construction schedule established by the developer, meetings were held with the structural engineer (hereafter called Haynes) wherein the developer carefully limited the engineering effort of the structural engineer, requiring constraints with respect to the size and weight of members to be used in the structural steel system. Additionally, the developer demanded that the type and slope of the multideck structure be dictated according to prior plans for a single-deck stadium which it had previously designed on an earlier project.

Before construction bids were let, the structural drawings were prepared by Haynes and delivered by the developer to the city. These drawings were turned over with full knowledge by Jamison, Haynes, and the developer that no final checking of the calculations had been performed and that an engineer's seal had not been placed upon them. All of this was the result of the fact that the developer had not yet reviewed and approved the drawings. Nevertheless, a construction committee appointed by the city reviewed the unchecked plans and accepted them.

At this point, a 1-year delay in the project was encountered because the city ran into legal difficulties in securing the bond issue which was to finance construction of the stadium. After the 1-year moratorium, the developer demanded increases in the cost of the project, which demand was unable to be met by the city. An agreement was worked out whereby the developer turned over the unsealed plans to the city and was permitted to walk away from the project without further involvement.

Some time later, bids were let on the construction portion of the project and a contract was awarded to a local contractor. Before signing the construction contract, the contractor insisted that the seal of the engineer be placed upon the plans. However, the engineers from Haynes who had prepared the unchecked drawings were out of town. A call was placed by George Rowe, project manager for Jamison, who advised the representatives of Haynes that their seal had to be placed on the plans that day in order to finalize the construction contract. Permission was given by the Haynes engineers for Rowe to break into the desk where the seal of Haynes was located so that it could be placed upon the drawings to satisfy the demand of the contractor. This was done with full knowledge by Rowe that the plans were unchecked at that time. A week later, all the plans were signed by the engineer from Haynes.

During the difficulties over the bond issue, Jamison encountered financial problems and went into bankruptcy. Rowe, an employee of Jamison, thereupon entered into a contract with the city whereby he would assume responsibility as the project architect during construction. When the construction contract was let, and the plans subsequently sealed by Haynes, Rowe entered into a construction-review contract with Haynes whereby the latter was to review the detailed drawings of the steel fabricator during construction. This review of the shop drawings did not contemplate a review of the calculations which had formed the basis for the structural steel plans.

Several months later, while in the process of reviewing the shop drawings of the fabricator, Rowe terminated the Haynes construction-review contract and retained Smee, a structural engineer with a one-person office. Smee was later to testify that he never conducted any review of the original unchecked drawings but merely proceeded to review shop drawings of the fabricator.

Accordingly, the project then proceeded through construction with a set of structural plans which were unchecked and which were based on preliminary calculations, with a design which had been dictated by a tight budget constraint with regard to the size and amount of the steel members to be used, and with supervision by a structural engineer who had not previously checked the adequacy of the drawings being used for construction.

After placement of the superstructure was completed, the contractor noticed the buckling of several structural support beams beneath the concrete decking. Investigation was thereupon commenced by independent experts, who verified the inadequacy of the structural design and recommended that immediate remedial measures be taken to bring the stadium up to design standards.

An action was commenced demanding damages of $1.2 million citing both the inadequacies in the structural steel system as well as various other architectural defects, such as poor sight lines which precluded spectators from seeing a good portion of the playing field. Named as defendants in this action were the turnkey developer; Jamison, the now defunct engineering joint venture; Haynes, the structural engineer; Rowe, the project architect; and Smee, the successor engineer to Haynes.

Testimony was taken from representatives of the turnkey developer, who acknowledged that all structural design work and schematic drawings had been furnished from its office to Jamison and Haynes. The strict guidelines of the developer dictated that no plans could be forwarded to the city without the express approval of the developer. In fact, one change to the drawings by Haynes calling for increased steel was rejected by the developer, who demanded that it be revised to avoid overdesign of the project in order to keep costs down and to stay within the original construction budget.

Testimony was also taken from the engineer from Haynes and from other officials of the company with respect to the circumstances surrounding the placement of the engineer's seal on the drawings with full knowledge that they had not received a final check. The Haynes representatives acknowledged that the placement of a seal upon plans normally represents that (1) the documents are correct as sealed, (2) an orderly process of inspection and review of the structural plans has taken place, and (3), assuming that the client had not been informed otherwise, the plans were checked before they were signed and sealed.

Finally, testimony was adduced that the project architect, Rowe, was aware that the plans had not received a final check when he received permission to break into the drawer of Haynes to remove the seal and place it upon the plans before the construction contract was signed.

Needless to say, extensive expert testimony was secured by all parties. All experts concurred that there was a total absence of necessary cross-bracing. Thus, the question ultimately reduced itself to a means of determining the proportional liability of the various defendants for their part in permitting the unchecked plans and drawings to go through the construction stage. Fortunately, the defective design was detected prior to completion of the stadium, and repairs could be effected without danger to person or property.

After extensive analysis of the role of each of the parties through discovery, and analysis of the various documents detailing the costs of repairs which were necessary, a settlement of this litigation was effected prior to trial. Contributions by the parties were affected by the fact that Jamison was defunct, the turnkey developer had been bought out by another company, and Smee was a sole practitioner without insurance coverage. Nevertheless, the turnkey developer paid 15 percent of the total settlement of $900,000; approximately 60 percent was paid on behalf of Haynes, the structural engineer; 20 percent of the contribution was paid by Rowe, the project architect; and 5 percent was paid on behalf of Jamison.

TECHNICAL ANALYSIS

The deficiencies identified in this design included the following:

1. Refer to Figure 21-1:
   a. *Member No. 1:* This column is subject to bending as well as direct stress and was of insufficient size to resist the combination. There is some indication that the designer proportioned for direct load only. Moment in the column results from the difference in span length between the overhangs and the main span (C to D).
   b. There were no filler beams under the walls of the vomitories, rest rooms, and other entities. It appears that the designer may have designed for uniform loading conditions only, neglecting these local concentrations.

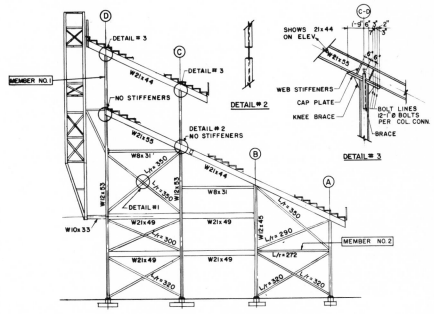

FIG. 21-1 Design of stadium.

c. Detail No. 1:   The design computations clearly show the assumption of $K = 1.0$ and L_u (unbraced length) equal to *half* the diagonal length in design of the diagonal bracing system. A counter system (i.e., compression diagonals buckle and become inactive) was assumed. Design was predicated on allowable $L/r = 300$ (maximum permitted for tension braces, other than rods). Apparently, at least some of the diagonal braces were not connected at their intersection. The result is that L/r values (based on $K = 1.0$) were greater than 500. Such a condition is not per se a structural deficiency. However, if the braces are not tightly fitted (no "draw"), or are not precisely straight when installed, lateral displacement in excess of the elastic displacement can be expected with a "counter" system.

d. Detail No. 2:   Note the lack of stiffeners. Lateral rotation (twist) of a number of these beams was observed.

e. Detail No. 3:   This drawing was copied from the corresponding sketch in the design computations. For convenience, the designer sketched only half of the detail, noting "12 1-in-dia. bolts, 6 each side of the column." As drawn, it would be possible for the detailer to have provided 12 bolts (3 lines of 4 bolts per line) on one side of the column. Fortunately, the flange width of the girder only permits two bolts per line, so that such a misinterpretation could not result—but this is not the best drafting technique. Also note that the girder is indicated as a W21x55 in the detail and a W21x44 in the section—another indication of hurried completion of the drawings. Finally, this detail is not adequately strong to resist (at normal levels) the moments resulting from wind and lateral live loads.

f. Member No. 2: L/r is 272 for this member, which clearly must function in compression. The designer appears to have proportioned it as a tension brace.

g. Note that there are no horizontal struts at the level of the tops of the footing pedestals. The horizontal component of the force in the diagonal braces (particularly after the compression diagonal buckles, as is intended in a counter system) must be resisted by tipping and passive soil pressure on the footings and pedestals. Movement is required to develop both the tipping and passive resistance, i.e., the soil must compress before it develops resistance.

2. Refer to Figure 21-2:

a. This structure acts as a rigid frame. Moments are induced in the rafter-column connections due to vertical, as well as lateral load (sway due to live load surge or wind load). The connection is a simple cap plate, bolted to the rafters, and provides little capacity for moment transfer. A similar connection is provided at the connection of column base plates to the foundation. The equivalent struc-

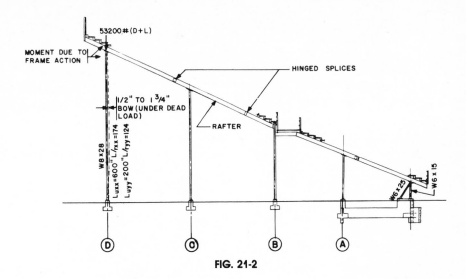

FIG. 21-2

tural action is represented in Figure 21-3. This structure is stable, i.e., a collapse mechanism does not exist, but is exceedingly flexible.

    b. The D-line columns were observed to have bowed as much as 1¾ in under deal load alone. This was attributed to neglect of the moment induced in the columns by the frame action. From a review of the design computations, this appears to have been the case. Note that L/r about the strong axis was critical for this member. A question was raised whether the designer might have overlooked this fact. The question was not resolved.

3. In general, review of the design computations indicates the following:

    a. The designer forgot to consider sidesway when estimating KL for the columns supporting the upper tier stands.

    b. In some cases, the designer incorrectly used r_{xx} (or r_{yy}) rather than r_{zz} in designing angle braces.

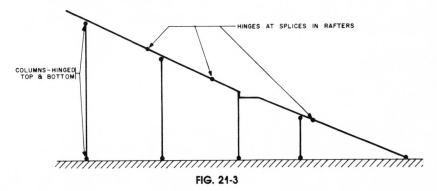

FIG. 21-3

c. The buckling of some of the bracing struts cannot be explained by a theoretical design analysis, since they are intended to act in tension. After installation, however, compression forces appear to have been induced by temperature effects and by fabrication and erection tolerances, which caused the exceedingly flexible struts (L/r of 300 or greater) to buckle.

Review of the computations indicates that the structural designer of this project was experienced. The computations show a reasonable level of technical capability. How could such an incredible mélange of basic errors have occurred? In this case, the evidence suggests two causes: (1) the lack of adequate time and (2) the engineer was not strong enough or sufficiently sure of himself to stand up to the pressures exerted by the developer—a self-styled expert. It is evident that the design and the plans were rushed to completion. This lack of time, of course, precluded the design and the plans being checked.

This case provides several insights into engineering practice which should be noted by both owners and engineers.

1. The owners contracted for the work with a developer who became the "boss of the job." The developer and the owner had negotiated a fixed price for the work. The developer's basic concern was to bring the job in at, or below, this price. It is understandable that he exerted a strong influence on the designers, who were working for him. However, such an arrangement (owner-developer) is common and a good enough arrangement if the developer is fully conversant with a project of this nature. In this case the developer, either knowingly or in ignorance, represented himself as an expert in the design of facilities of the type contemplated. He was not! As a result, he underestimated the cost of the project. Faced with a potential loss, all manner of questionable actions might seem justified.

2. The owner is not blameless in this case. Why were the qualifications of the developer not better checked before awarding him a contract? Why did the owner not retain the architect-engineer, rather than leave the selection to the developer? Surely, the owner was aware that this action bypassed the normal set of checks and balances operative in the evolution of a construction project. In lieu of such procedure, did the work of the owner's "construction committee" include an independent review, by qualified professionals, of the developer's plans?

3. Finally, the above listing of the deficiencies which were found in the structure constitutes an excellent checklist of common errors encountered in structural design.

LEGAL ANALYSIS

This case evidenced all of the legal complexities which can confront counsel seeking to represent design professionals. There was no doubt from the outset that a substantial design defect had been encountered, nor did any party fail readily to discern the fact that the overlap of functions and a noticeable lack of coordination between the design and its construction did not bode well for the project.

Clearly, the contractual relations in this case placed an inordinate amount of restrictions upon the latitude of the engineer. One is tempted to ask how the engineer could permit himself to be placed in such design handcuffs that he would permit an underdesigned, unchecked set of drawings to leave his office. During the scope of discovery, it was learned that the engineer vested with overall responsibility for the design in the structural engineer's office had no prior experience in designing a multideck stadium. Consequently, he was most deferential and resigned to accept the design criteria established by the turnkey developer, who, in turn, presented himself as having extensive experience in designing stadiums of this type. This, of course, proved to be untrue.

Responsibility for the malfunctioning of the normal design-construction process on this project also must rest with those who orchestrated the project on behalf of the city. Certainly no one was in control of the entire process so as to coordinate various design trades. It cannot be pointed out strongly enough that a design conceived by a developer, who contemplates constructing the project from his own plans, will take an entirely different direction from that designed by an independent engineer who expects to supervise construction of a general contractor operating in accordance with a separate contract for construction. Placing entire control of the design and construction of a project within one entity may help to keep costs down. However, there is a good possibility that inbreeding along such lines will preclude review and analysis by an independent engineer or contractor with the attendant benefit of bringing a fresh eye to the project.

Thus, counsel was presented in this action with a project which required analysis of every step of the design-construction process. Following the ebb and flow of the design drawings as the process unfolded represented a Rube Goldberg invention. It cannot be stressed enough that architects and engineers should avoid at all costs any type of situation in which control of design is wrested from their hands by another only to find that responsibility for sealing the plans must fall on their shoulders.

Once again, the issue of damages was critical to the design professionals as a precondition to determine any apportionment of liability. In this instance, the project architect and the structural engineer agreed to pool their resources toward the retention of a single structural engineering expert who analyzed the remedial work which had been performed on the part of the city. This analysis led to a conclusion that the remedial work incorporated certain design features which went beyond the original scope of the project, creating a betterment to the city. Therefore, it was argued that a certain portion of the remedial costs could not be chargeable to the defendants since they increased the scope of the project beyond the original plans and drawings. Attention to facets of a damage claim such as this will enable counsel to narrow the scope of provable damages and assist in assessing the realistic exposure at an early stage in the case.

case 22

TYPE OF FACILITY
Power plant (fossil fuel)

TYPE OF PROBLEM
Excessive differential settlement

Significant Factors

A. Dissolution of gypsum in underlying rock, creating cavities which collapsed

B. Failure to relate significance of foundation conditions (gypsum) to use and operation of the facility (infiltration to ground of heated discharge water)

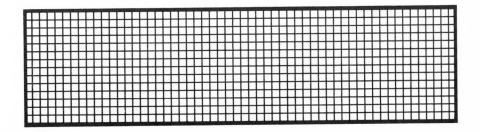

NARRATIVE

This project involved the construction of Units 4 and 5 of a fossil fuel power plant in a southwestern state. The foundations for the units included, in part, spread footings bearing on an upper stratum of sandstone and shale and, in part (for heavier loads), belled caissons (see Figure 22-1) bearing on a lower stratum of sandstone.

Shortly after the plant was put into operation, distortion of the steel framing in the superstructure was noted. Beams were observed to be twisted. Braces were buckled. As time passed, the bridge cranes began to have difficulty in operating. The crane rails (and runways) had sagged. Finally the turbine bases began to show a perceptible tilt. Investigation revealed that differential settlements of as much as 1½ in had occurred. The cause was traced to the presence of lenses of gypsum in the upper sandstone-shale stratum, under the spread footings. Operation of the plant involved the use of cooling water from a nearby cooling pond. Water seepage from the pond and from the intake and discharge canals had caused a rise in groundwater level. Further, since the discharge water was warm, the groundwater temperature was raised. Both effects caused the gypsum

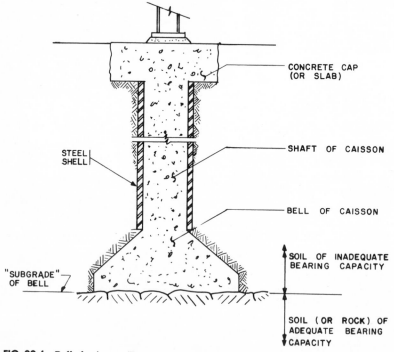

FIG. 22-1 Belled caisson. (Excavation, normally, is by power auger, with final trimming of bottom by hand.)

to dissolve, creating voids in the ground below the shallow foundations. These voids progressively collapsed under the weight of the structure. The caisson foundations which bore on the lower sandstone layer—which had a lesser or no gypsum content and was less permeable than the upper layer, so that the circulation of groundwater was impeded—did not settle.

Repair was accomplished by "mudjacking."[1] A soil-cement mixture was injected at a pressure of 200 lb/in² and the sagged units were raised back to proper elevation. Interestingly, the turbine pedestals (mats) were not mudjacked, only the superstructure supports. The reason was that the releveling would have required disconnecting the various pipes to the turbines, their realignment, and complete overhaul. Further, it was felt that the stiffness of the pedestal mat would have caused unequal stress to exist in the soil after the mudjacking was completed; the stress would gradually equalize (iron out), causing a new round of differential settlement. The stiff mats had tilted but remained plane and the turbines remained operable.

TECHNICAL ANALYSIS

The case, as described in the Narrative, is simple. The following technical points are noted.

1. The presence of gypsum lenses in the upper sandstone-shale layer was known during the design stage—but the significance was not realized. The soils engineer did not relate the operation of the plant, infiltration from the intake and discharge canals, the cooling pond, and the elevated temperature of the discharge water to his problem, which was to recommend the type of foundation and the allowable contact bearing pressure. His defense, in essence, was the following: How was he to expect a rise in groundwater level? How was he to know that the canals were not lined, thereby increasing the amount of seepage into the ground water? How was he to know that the influent water would be heated? This is a valid argument, but the fact is that a foundation is not an entity nor an end in itself. It must be "right" for whatever it supports and it may be that, for like conditions of soil profile and of loading, the foundation that is right for a factory building may not be right for a public building, or for a power plant, or for a paper mill. Each has a different tolerance to settlement. The author differentiates soils engineers and foundation engineers based on this ability to discern differences in type of required support, depending on the use of the structure and the type of framing. Ideally, the structural designer and the founda-

---

[1] A procedure wherein a soil or soil-cement slurry is injected under the foundation at a pressure sufficient to overcome the weight of the foundation and thus raise it. The injection also fills voids in the ground under the foundation.

tion designer should be the same (or each should have both capabilities) so that a proper marriage of the two elements can be realized.

2. As an example of the principle described in the preceding paragraph, note that the turbine pedestals were *not* releveled, despite differential settlement. So long as they retained a plane surface they remained operable.

3. The matter of locked-in stresses in the steel framing of the superstructure (due to the settlement and remaining after releveling) was investigated. The question was whether or not connections had to be disassembled to relieve restraints and so to relieve the locked-in stresses. It was concluded that no important damage had been done and that relief of restraints was not required.

4. The reason for the mixed foundation (spread footings and belled caissons) was the need for uplift resistance in part of the structure. Belled caissons were used where uplift capacity was required. The less costly spread footings were used elsewhere.

5. The upper sandstone-shale layer was partly weathered and was more permeable than the lower stratum. Contact bearing pressure on the upper stratum was established as 15 kips/ft² and on the lower stratum 30 kips/ft². The inclusions of gypsum were a fraction of an inch thick and were found in the upper stratum, mostly above, but in two instances below, the groundwater level. The gypsum lenses showed some evidence of a honeycomb structure, which accelerated the percolation of water and the resulting dissolution of the gypsum. This fact was not known prior to design of the foundations.

6. Units 1, 2, and 3 of the power plant, built sometime prior to Units 4 and 5, were founded, entirely, on the lower sandstone layer and did not experience differential settlement. One may not lightly discount previous experience or the judgment of others.

LEGAL ANALYSIS

The damages sustained by the utility were far in excess of the soil consultant's financial ability to pay, and therefore the utility sought to recover not only from the soils engineer but from the property insurers who provided the utility with coverage for property damage. The property insurers contended that the damage was due to an error or mistake in design, and therefore coverage would not be afforded by virtue of an exclusion contained in their policy. When the insurers declined to honor the utility's claim, the utility commenced a suit against its insurers and against the soil consultant. Just prior to trial, counsel moved to sever the action against the soils consultant and to try this action separately from the action against the insurers. The utility, in an opposing motion, argued that the question to be determined was whether the settlement was due to an error or defect in the plans and specifications. The utility argued that if no error or defect was found, the property insurers would be held liable to make payment under the provisions of the policy and the soils engineer would be held not liable. Conversely, if an error or defect were determined to be the cause of the settlement, the exclusion in the policy would be applicable and, while the insurers would escape liability, the soils engineer would be found liable. The argument on its face appears logical; however, there is an important distinction between error or defect and negligence. While negligence involves an error or defect, not all errors or defects constitute negligence. To constitute negligence, there must be a showing that the design professional did not act in accordance with the standards of the profession. If he acted within the standard but there was an error nonetheless, there is no negligence and no liability.

The court accepted this basic principle and granted the motion to sever the consultant's trial. The utility was now on the horns of a dilemma. If it proceeded first against the soils consultant and proved negligence, there would be a finding of a defect or error which would preclude it from proceeding against the insurers. While it would thereby make a recovery, that recovery (due to the financial limitations of the consultant) would be less than 10 percent of the amount it could conceivably recover from the insurers.

On the other hand, if it proceeded first against the insurer, it would have to argue that there was no defect or error in the design of the facilities. This position would be inconsistent with the position that the utility would advance against the engineer.

The trial court found that there had been an error or a defect in the design; and that therefore the property insurers had rightfully declined payment on the loss. The matter was sustained on appeal, and the utility company thereafter proceeded to trial against the soils consultant.

The primary defense available to the soils consultant was lack of negligence. The fact that an error had been committed in not recognizing that the gypsum was honeycombed and, therefore, readily susceptible to dissolution was not subject to denial. The defense, rather, would be based upon the fact that standard soil testing procedures resulted in a crushing of the gypsum during the taking of the samples. As a result, an analysis of the sample, while revealing a gypsum layer, would not indicate its condition in its natural state. The defense would be buttressed by the utility's own expert who had testified in the trial against the insurers. The utility, in attempting to put its best case forward against the more financially viable insurers, had offered expert testimony that there was no error or defect. While the testimony was not persuasive on the issue of error or defect, it had a direct bearing on the issue of negligence; the utility now was hard pressed to explain away the prior testimony it had offered.

The simple distinction between error and negligence and the stratagem of severing the trials to focus on that issue had set up the matter so that a settlement on behalf of the consultant was achieved short of trial and within his capability to pay. While the utility did recover monies, it amounted to less than 10 percent of the actual damages.

case 23

TYPE OF FACILITY Monumental public building

TYPE OF PROBLEM Excessive differential settlement

Significant Factors

A. Failure to set foundations on the intended bearing stratum

B. Inability of the construction inspection staff to evaluate the adequacy of the soil strata encountered and, therefore, to set the foundations on material of adequate capacity

C. Disturbance of the subgrade prior to, or during, placement of the concrete for the foundations

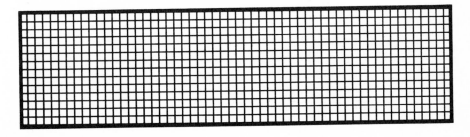

NARRATIVE

The foundation design for this building, as originally proposed, called for belled caissons (see Figure 22-1). Caissons supporting the main structure were proportioned for a contact bearing pressure of 40 tons/ft² and were to bear on sound rock (mica schist) at elevation −70. There were 186 such caissons. Caissons supporting ancillary structures were proportioned for a contact bearing pressure of 8 tons/ft² and were to bear at a higher level. There were 124 of these.

The contract, as issued for construction, recognized that sound bearing material would not be found at a uniform depth or elevation and gave the discretion for determining the actual founding elevation for the caissons to the resident engineer. To assist him in this judgment, the contract provided for testing the proposed subgrade for the bells of the caissons to determine if the soil (or rock) material was of adequate capacity to support the design contact pressure.

The proposed test was the simple Standard Penetration Test (SPT), which is used when making borings. The test consists of driving a standard size tube, using a drop hammer of standard weight and fall. The resistance to penetration (blows/ft) is a measure of the density and, therefore, strength of soil. A resistance of 150 blows/ft was specified as the desired touchstone.

The contract was later amended (before bidding) at the behest of the owner to require more sophisticated testing of the subgrade. Samples (2-in by 2-in section, by 4 in long) were to be cut from the material at subgrade and tested for compressive strength (unconfined compression test). Proper attention was to be paid in the testing to orienting the sample so that the foliations in the sample were in the same position under the head of the testing machine that they would be under the bell of the caisson. The touchstone was to be that material which tested for 160 tons/ft² or greater would be acceptable for support of the proposed contact pressure of 40 tons/ft². Recognizing that the recovery of such samples, the testing, and the reporting would be costly and time-consuming, test samples were to be taken only in a portion of the caissons. Elsewhere, reliance was to be placed on a form of penetration resistance test. Under each caisson, the contractor was to drill into the proposed subgrade to a depth of 5 ft, using a specified model pneumatic drill with a bit of fixed diameter. The idea was to record the time it took to drill the 5 ft, to correlate that time to the laboratory strength tests, and so to derive the time it would take to drill 5 ft (using a drill of fixed size, powered by equipment of fixed output) through rock which corresponded to a strength of rock of 160 tons/ft², i.e., adequate to support the proposed contact bearing pressure. In essence, it was a form of SPT, but using rate of penetration instead of impact resistance as the weighing factor. The drill test holes were to be filled with grout before placing the concrete in the caisson bells.

The owner was a governmental agency. It was represented on the site by a resident engineer. Working under the resident engineer, and acting under contract to the government, was a testing laboratory. The laboratory provided a "qualified caisson inspector" to observe the field operations. The timed-drilling tests and the recovery of the specimens for laboratory testing was done by the contractor under inspection by the caisson inspector. Laboratory testing was done by the testing firm. Results of the laboratory tests and of the timed-drilling tests were reported to the resident engineer, who then decided if the caissons had to penetrate deeper or could be founded at the test elevation. Thus, the decision of the owner's representative was based upon information presented by the testing laboratory.

As construction of the building advanced, cracking of the exterior walls was observed. Investigation revealed that up to 1½ in of settlement of some of the interior caissons (40 tons/ft² contact pressure) had occurred. Further investigation showed the following:

1. Caissons proportioned for 8 tons/ft² contact pressure had not settled.

2. Caissons had been founded, on average, 18 ft higher than the original design level of elevation 70.

3. The caissons did not bear on sound rock, but on decomposed rock (mica schist). The decomposed rock did not have adequate strength to support a contact pressure of 40 tons/ft² and had, indeed, settled under the lesser pressure which developed during construction.

Certain peripheral causes, such as defective concrete in some of the bells and shafts were also uncovered, but the proximate cause of the observed settlements clearly was the inadequacy of the subgrade material to support the more heavily loaded caissons. Further investigation centered on why the caissons had not been founded on sound material (rock), and on who was to blame. An action was started by the owner against the architect, structural engineer, testing laboratory, general contractor, and various subcontractors, seeking $5 million.

The question was raised whether the testing laboratory had provided valid information, and what responsibilities were to be borne by the contractor, his several subcontractors, and the architect-engineer.

The owner alleged that it had depended upon the testing laboratory to advise that the material at the bottom of the caisson excavation was suitable for placement of the caissons. The complaint asserted that the laboratory misled the owner in that the material at the bottom of the caisson in fact consisted of decomposed mica schist and was not suitable for the foundation as designed.

During the course of giving his initial deposition, the construction manager (employed by the owner) admitted that the final responsibility for determining whether the excavation had been carried to a sufficient depth was his alone, and was dependent upon various factors, only one of which was information furnished by the testing laboratory. He acknowledged that these additional factors included the results of the unconfined compression tests on the rock. Those tests had been prepared by the laboratory who also advised in regard to the results of the drilling tests. However, he also admitted that many of the caissons were poured prior to the results of the compression tests being made available to him. Apparently, this was done to save time and money.

On behalf of the laboratory, it was argued that the mica schist may not have been decomposed at the time the tests were run. There was testimony to indicate that water was present at the bottom of some of the caisson excavations and, according to expert testimony, water often can lead to decomposition (slaking) of mica schist. It also was pointed out that during construction of the caissons, disturbance of the underlying mica schist by the placement of concrete and the trampling by the workers could have lead to decomposition of the mica schist.

The case, though litigated, never went to the point of final determination. Alleged damages of $5.9 million were settled for $750,000 ($200,000 contributed by the architect-engineer and $550,000 contributed by the contractor group). Settlement was made without admission of fault, but as a lesser evil to the costs of continuing litigation and the vagaries of a court decision.

Remedial measures to avoid further settlement involved pumping grout under pressure below the bases of the affected caissons, thus cementing and strengthening the soil.

TECHNICAL ANALYSIS

Significant factors in this case include the following:

1. Three types of foundation design were developed for the project and were included as alternatives in the contract—steel H piling, pressure injected footings (Franki piles), and belled caissons. The belled caissons were bid about $0.5 million less than the other alternatives. It is interesting that the architect-engineer recommended against the use of caissons. If the alleged damages of $5.9 million were true and the owner settled for $750,000, perhaps the architect-engineer was right. One wonders what formed the basis of his objections. Had he some experience in the area, with the soils in question? The situation at the site did invite problems. There was difficulty in keeping the holes for the caissons dry. In fact, the resident engineer permitted 166 of the caissons to be filled with concrete before the applicable laboratory test results were received, apparently in an effort to stay ahead of the water. He relied

on the timed-drilling tests only. The combination of air and water on an unconfined, exposed surface of decomposed mica schist is bound to result in some slaking.

2. If one would use a caisson foundation where inspection of the subgrade is required, one must recognize the practical difficulties of performing that inspection. In this case, the shafts were 30 in. in diameter, about 70 ft deep. The resident engineer (give him due credit) wanted to enter some caissons to see for himself what he was approving. He was dissuaded from doing so—and properly. Inspection was done by a junior (we hope young and *slim*) man. Investigation revealed that even the official "caisson inspector" did not enter more than one-fourth of the caissons. He had to stay topside to monitor the pressure on the pneumatic drill during the timed-drilling tests. Apparently, by the time the drill operator went into the hole, did the test, and came out there was enough water in the bottom of many caissons that the inspector couldn't have seen much, even if he did enter.

3. The problem of disturbance of a wet subgrade due to placement of concrete thereon and by the feet of the workers is classic. Something of that sort likely occurred in this case.

4. The caisson inspector was not a geotechnical engineer. Neither was the resident engineer. Apparently, decision on this sensitive geotechnical evaluation was to be made without having anyone with that background look at the soil *in situ*. There are many cases where an experienced layperson knows more about a problem than a person with technical training—but it is a risk. In this case, the decomposed schist could be crumbled by hand. It was really a granular material, rather than a rock. Failure to appreciate the difference was surprising.

5. The laboratory compression tests, instead of being performed on 2-in by 2-in by 4-in specimens, were performed on 2-in cubes. Apparently, little or no attention was paid to the orientation of the foliations in performing the tests. Subsequent testing indicated that the results were not sensitive to these two discrepancies from the specified procedures, although one would think that they should have had some effect. A curiosity was that all the timed-drilling tests on over 300 caissons were recorded to penetrate the specified 5 ft in about the same amount of time. These facts raised questions as to the validity of the test results which were reported. As noted, the matter was never resolved.

LEGAL ANALYSIS

Following the commencement of litigation, the initial meetings between counsel for the parties indicated that several hundred thousand documents would be subject to review and circulation among all parties. Additionally, it was anticipated that as many as sixty depositions would be required before going to trial.

However, immediately after the responsive pleadings were filed by the defendants, the deposition of the owner's construction manager was noticed. Prior to this deposition, the construction manager had been required to review over 100,000 documents in order to answer specific written interrogatories which had been served by the defendants upon the owner. Consequently, the deposition of the construction manager took more than 20 days and involved intensive questioning about the manner in which he proceeded and the specific steps he took with regard to the design, foundation placement, and settlement problem.

As a result of this deposition, defense counsel were of the opinion that sufficient weaknesses existed in the owner's case to preclude a confident sense of recovery on the part of the owner. This belief was further supported by evidence which tended to indicate that the owner's on-site representatives may have lacked sufficient expertise in their own right to make the necessary decisions with regard to the foundation design. Consequently, pressure was brought to bear upon the owner to determine whether he wished to proceed with the full discovery procedure, as well as the time and expense associated with pursuing this matter to trial.

Although the owner initially expressed no desire to settle for less than a multi-million-dollar sum, following the completion of the construction manager's deposition, these negotiations did commence and ultimately resulted in a settlement that clearly reflected "costs of defense" to the defendants.

In reviewing a case of this complexity, where the details of the particular testing procedures and decision making associated with them were pointedly at issue, counsel must be well-versed in all of the nuances of foundation design and construction. Such intimate knowledge enabled defense counsel to pinpoint questions to the construction manager so as to secure admissions from him that the ultimate decision for determining whether the excavation had been carried to a sufficient depth was, in fact, made by him. This turned out to be at variance with the complaint and cast doubt upon the success of the owner's case had the matter proceeded to trial.

case 24

TYPE OF FACILITY   University building

TYPE OF PROBLEM   Injury to a worker

Significant Factors   Third-Party suit in which the architect-engineer
                         was held liable for injuries to workers; which
                         injuries occurred during erection of the
                         building

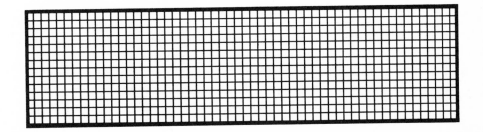

NARRATIVE

The action in this case arose out of an accident which occurred during erection of the steel framework for a building at a midwestern university. At the time of the occurrence, two ironworkers were erecting steel at a level approximately 43 ft above the concrete floor of the building. The workers were attempting to connect a horizontal steel beam to the web of a vertical steel column. Another beam had previously been connected to the opposite side of the column in what is commonly known as a "double connection," i.e., both beams are connected to the common web (of a column or girder) using the same bolts (see Figure 24-1). In making such a connection, one beam is erected first and is supported by the connecting bolts. To connect the second beam, it is necessary to remove the nuts on the bolts and, perhaps, some of the bolts themselves.

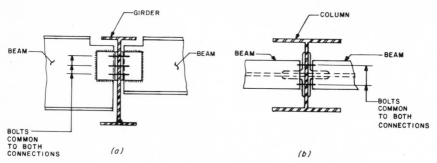

FIG. 24-1 Double connection. (*a*) To a girder. (*b*) To a column.

At the time he was attempting to make the connection, one of the workers was sitting atop the beam that had already been attached. The other worker was sitting on a joist which framed onto the same beam, near the connection. The beam to be attached had been lifted into position and was being held by a crane. One of the workers held the beam in position while the other removed the nuts. At this point, it appears that the column was totally unsupported (no wires, no bracing) except by the anchor bolts at base slab level, 43 ft below. For whatever reason, the top of the column began to move aside and the beam which was first attached began to separate from the column, withdrawing the bolts which no longer were restrained by the nuts. The beam, the joist, and the workers fell to the concrete. Both workers suffered serious personal injuries.

A suit was thereafter commenced by the injured workers against the general contractor for the project, alleging negligence in connection with the accident. The contractor filed a third-party complaint against the employer of the workers, who was the subcontractor for the fabrication and erection of the steel. The contractor also commenced a third-party action against the architect for the project. Later, the injured workers also joined the architect as an original defendant in this action.

A trial of several weeks was held with extensive testimony concerning the connection methods employed in joining the columns and beams on this project, the applicable requirements set forth by the AISC, the identity and responsibility of the parties chargeable with the shop-drawing detail for the connection, as well as the relative liability among the parties for the injuries sustained.

On behalf of the architect, it was acknowledged that he had prepared the plans and specifications, but methods of erection and safety procedures were alleged to be the responsibility of the steel erector. The architect, after preparing the design, had no responsibility for the method by which the building was constructed and thus asserted that he had no control over the safety measures taken during construction. It was conceded that the architect had a duty of care with respect to supervision and design and had, in fact, approved the connection specified by the fabricator. However, the architect argued that it was the fabricator who had the necessary expertise to design the connection.

At the trial, expert testimony was adduced on behalf of the workers to the following effect:

1. The "slenderness ratio" (L/r) of the steel column exceeded the maximum ratio set forth in the AISC (steel buildings) design manual. The column was deemed excessively flexible.

2. Connection methods other than the "double connection" which was used herein, should have been used for framing located more than 15 ft above ground.

3. The original shop drawings had not shown a double connection. The detail had been changed as a result of the architect's review.

4. A prior, almost identical, accident had occurred on the job approximately 6 weeks before this accident. The architect knew of that accident and did nothing to prevent a recurrence.

5. The architect's contract with the owner was to "supervise and coordinate at the site the work of several contractors. . . ."

The jury returned a verdict releasing all parties from liability, except the architect. The jury found the architect liable to the injured workers and their families in the sum of $830,000. Upon appeal, the verdict of the jury was upheld.

TECHNICAL ANALYSIS

There is no technical analysis in this case. There is no technical problem. The case is included as representative of a burgeoning class of actions being directed against architects and engineers.

LEGAL ANALYSIS

A brief legal analysis of this case cannot do justice to the complicated questions of fact and law which were presented in the matter. The relative liability of the parties, the varying nature of the experts retained by the various parties, testimony in regard to whether or not specific standards were applicable to this method of connection, and the question about the extent of the injuries sustained by the two workers were all intensively litigated.

The jury undoubtedly was influenced by the nature of the accident. However, evidence was presented which proved that at the time of the trial one of the injured workers was working at a job from which he earned more than he had been making at the time of the accident. While the other injured worker had suffered fractures of various parts of his body, testimony at the trial showed that his own physician believed he could hold a sedentary job. The second injured worker testified that he had been familiar with the type of connection he had been making at the time of the accident, having done this task perhaps a thousand times before.

The jury also heard testimony from the worker who had been injured some 6 weeks prior to this accident. This witness testified about the similarity between the two accidents and furnished his own opinion that the two workers would not have been injured if a flange connection, a seat connection, a staggered connection, or a safety-hold connection had been employed by the designer. He testified that the double connection utilized was generally acceptable at a height of between 10 and 15 ft but never on a column as high as 43 ft.

The jury also heard from an expert engineer who was a member of the AISC and had served on the Manual and Text Book Committee of that organization. This expert testified on behalf of the architect that he was familiar with the double connection used at the time of the accident and that it complied with the standard in the profession. Other experts put on by the plaintiffs testified to the contrary with respect to this compliance issue.

There is no one essential lesson that can be learned from a case of this kind. Conflicting expert testimony will always present a question of fact to a jury; it is left to their discretion to decide which testimony is to be relied upon. All counsel appeared to express their utter amazement at the amount of the verdict although no one doubted that the workers were entitled to a reasonable recovery for their injuries. To the extent that the verdict was rendered solely against the architect, there appears to be no rational basis to conclude that the other parties to this action, i.e., the general contractor and the fabricator of the steel beams, were entirely without fault. There was sufficient testimony to enable the jury to find proportionate liability against these two parties but it chose not to make such a finding. Judges and appeal courts rarely will disturb such verdicts.

case 25

TYPE OF FACILITY   Warehouse

TYPE OF PROBLEM   Excessive flexibility and cracking of framed first floor (over basement)

Significant Factors   The structural engineer neglected to consider the loads imposed by the forklift trucks

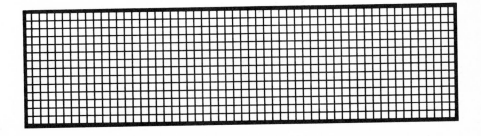

NARRATIVE

The subject in this case is a one-story warehouse with a basement. The problem relates to the first-floor structure over the basement. The construction is a 3-in concrete slab on open-web steel joists on rolled steel girders.

The structural elements of the warehouse were designed by a local engineer of proven experience, under contract to the general contractor. The engineer also was under contract to serve as the owner's professional representative for supervision during construction. Design was based on a uniform floor load of 200 lb/ft² in the finished product storage areas and 300 lb/ft² in the areas to be used for storage of raw materials. These load requirements were provided by the owner.

After the warehouse was in operation, the owner noted that the floor vibrated alarmingly when the forklift trucks used to handle pallets of the stored materials passed. This prompted an investigation which revealed that the designer had not considered the concentrated loads from forklift trucks (which each weighed over 5000 lb) when designing the floor system. He had designed only for the uniform load condition, i.e., the stored pallets. The resulting structure was substantially underdesigned and, to quote one expert, "totally unsuited for a warehouse."[1]

As a result of an inspection of the facility, four separate problems were found to exist:

1. Floor vibration as the forklift went from the existing building to the new addition. Steel plates joined the floors at this point.
2. Hairline cracks throughout the entire floor.
3. Deteriorating concrete at the joints, apparently attributable to the fact that the joints had not been filled with sealant as recommended by the engineer.
4. The load of raw material pallets, when added to the load of the forklifts, was in excess of the recommended loads for the joists.

The problem was corrected by removing the entire first floor and replacing it with a new floor, built by a stronger method of construction. Rolled steel beams were used in lieu of open-web joists.

An arbitration demand was filed seeking damages against the engineer of $94,000. Settlement with the owner was effected for $83,000, assessed against the engineer.

The basic questions raised in this litigation related to the matters of (1) the obligation of the engineer to inquire into the intended use of the warehouse and (2) if the owner neglected, or failed, to specify a need

---

[1] Investigation also indicated that the owner had contributed to the problem by overloading the floor by as much as 30 percent.

for more than the uniform load equivalent, was the engineer expected to be sufficiently versed in warehouse operation to know that some form of concentrated loads due to material-handling requirements are a normal consideration in the design. The engineer's position was vitiated in this case by the fact that the structure was an addition to an existing warehouse, which he had visited several times, and where forklift trucks were used to move the materials into and out of storage. However, the answer to both points seems to be that an engineer does bear an obligation to acquaint himself sufficiently with the proposed use of the structure in order to establish realistic design loadings.

TECHNICAL ANALYSIS

The framing plan of the floor is shown in Figure 25-1. Floor construction consisted of a 3-in concrete slab on Corruform (corrugated metal form)

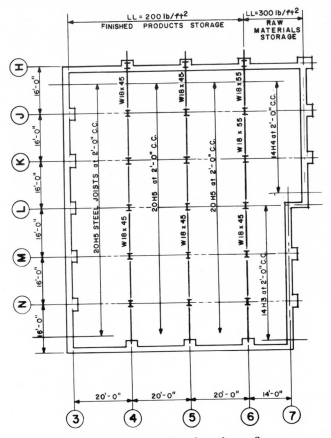

FIG. 25-1 Framing plan of warehouse floor.

supported on open-web steel joists on rolled steel girders. Reinforcement in the slab was a single layer of 66-1010 welded wire mesh.

As noted in the Narrative, the structural designer simply neglected to consider that, in addition to the storage load, the floor had to support the concentrated wheel loads of forklift trucks used to move the stored materials.

This type of omission is not uncommon. No special comment is required. Of interest, however, is an analysis of the stresses in the structure which would be caused by the reactions from the forklift trucks. The trucks used in this case had a weight of 5750 lb and a lift capacity of 3000 lb. For such a machine, the wheel loads are about 4200 lb each, spaced about 30 in apart. Computation based on the provisions relating to distribution of wheel loads as presented in the AASHTO Standard Specifications for Highway Bridges and neglecting impact indicates an *overstress* in the slab of about 350 percent and in the joists of about 50 percent for flexure and 130 percent for shear. Despite these overstresses, the structure did not collapse. It was in use and it was functioning. The only signs of distress were local vibrations and deflections occurring during passing of the forklifts and flaking of the concrete at the joints in the floor, which joints had not been filled or armored to resist the impact of the wheels.

Why didn't the floor, which did not collapse, at least show signs of inelastic distortions? One can speculate about whether the forklift trucks were carrying less than a full load. Also the degree to which the wheel loads and the storage loads were concurrent on the joists (the above-noted computations assumed no concurrence for analysis of stresses in the floor slab and no concurrence in an 8-ft aisle width for analysis of stresses in the joists) is a matter of speculation.

Further, an overstress of 50 percent in the joists should not necessarily cause collapse since the expected safety factor is greater than 50 percent.[2] However, none of these speculations can account for strength in the slab ample to resist a 350 percent apparent overstress and for strength in the joists ample to resist a 130 percent overstress. It can be concluded that the provisions of AASHTO regarding distribution of wheel loads to concrete floors and to the supporting stringers are quite conservative. Evaluation of a number of cases using a grid analysis (and the computer) and load tests performed on prototype slab panels confirm this conclusion.

LEGAL ANALYSIS

Immediately after the problems were observed, it became necessary to undertake an analysis of the design. Independent experts were retained

[2] Adding 30 percent overload by the owner would increase the overstress to a level of alarm.

who confirmed that the design failed to meet the requisite standards for supporting the necessary loads.

Once satisfied that the engineer was liable, counsel then was required to concentrate on the question of damages. Working with the independent experts, it was determined that the warehouse, in its present condition, was usable but that repairs would have to be effected to the first floor slab. Requests were made to the independent consultant to determine the various alternative remedial schemes which would solve the problem, including consideration of an alternate model forklift truck of lighter weight.

A consultant retained by the owner concluded that the floor should be removed in its entirety and reconstructed. This approach would have entailed the loss of the use of the facility for approximately 1 month, as well as additional storage costs for removing the materials in the facility and placing them at another location.

Ultimately, both the experts for the owner and the engineer reviewed the alternative methods of remedial design, agreed on the necessary amount of materials and work required for the repair and reviewed a proposed statement of damages.

A revised estimate of damages was submitted by the owner indicating a total repair cost and loss of use during the repair period totaling $104,488. Disputing this on behalf of the engineer, counsel pointed out to the owner that the new design represented a "betterment," i.e., an improvement to the facility, a portion of which should be borne by the owner. The concept behind this theory is that the owner of the property is securing a better floor by virtue of the new design than was originally designed and constructed. Consequently, it was incumbent upon counsel to secure an estimate of the repair costs which were alleged to be an improvement over the original design, in an effort to secure a reduction in the amount of claimed damages.

Subsequently, negotiations with the owner did result in a reduction of the settlement demand to $83,000.

case 26

TYPE OF FACILITY   Shopping center

TYPE OF PROBLEM   Excessive settlement

Significant Factors

A. Failure to appreciate that consolidation of a SILT soil (ML), albeit of stiff consistency, can produce significant settlement

B. Borings penetrated to insufficient depth

C. Failure to discover compressible stratum of 1½- to 3-foot thickness in explorations

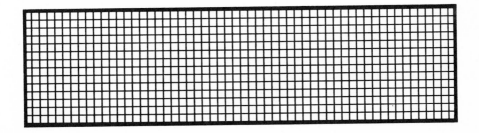

NARRATIVE

When owners encounter problems arising from the design and construction of a building, it is not unusual that they will seek to involve all parties who had anything to do with the construction process. If litigation ultimately ensues, this generally involves the joinder of all such individuals and entities to ensure that all possible theories of liability are asserted against the responsible parties.

However, problems sometimes arise when owners choose, for their own purposes, not to name certain parties who would otherwise be logical defendants in the action. When this happens, it often raises questions in the minds of the other defendants about the motive behind an owner choosing not to name a party who otherwise would be potentially liable. The common practice then is for the other defendants to join that nonincluded party into the action in order to ensure his or her participation in the litigation process and recovery from such a party for his or her share of responsibility.

The case at hand concerns the design and construction of a single-story department store. Foundation investigation proceeded in the usual manner. A firm of soils engineers was retained by the owner to perform subsurface investigations and soil testing and to prepare a soils report. The subsurface investigations (borings) were subcontracted to a well-known drilling company. The testing and the soils report were completed by the soils engineer. The soils report indicated that the site was the location of a 30-year-old fill, that the site grading required excavation at one end of the building and up to 30 ft or more of fill at the other end and that a conventional spread-footing foundation would suffice, but that the bearing area under the footings should be undercut a minimum of 4 ft and backfilled with select, compacted fill to form distribution pads under the footings. Contact bearing pressure not to exceed 4000 lb/ft² was advised. The soils report further advised that "slight differential settlement" should be expected and that vertical joints should be incorporated into the exterior masonry walls to accommodate the resulting distortions of the building.

The owner adopted the recommendations concerning the use of spread footings and maximum allowable contact bearing pressure. The new fill was placed under controlled conditions, to the stipulated density (within acceptable tolerance). In the main, the recommendations of the soils engineer were followed, except that the vertical movement joints were not incorporated in the masonry walls.

Shortly after construction was started, during erection of the roof, a crack was noted in one of the exterior masonry walls. At this time, the floor slab had not been poured. A second soils engineer was consulted. He advised the owner that the crack had occurred at the location of an old, abandoned sewer. It was thought probable that the crack was due to poor compaction of the old backfill in the sewer trench (an 18-ft-deep trench was dug to locate the sewer, but it was not found); that the settlement was a local, short-term phenomenon; and that the building could be completed without modification of the design. The wall, and some interior columns, were monitored for settlement for about a week's time. No substantive settlement was recorded. Accordingly, the building was completed and occupied.

About a year later, a major crack developed across the entire width of the floor slab of the building and through the exterior walls. The roof cracked and leaked. The building was being split; one entire end was settling, rotating about an axis corresponding to the location of the crack in the floor slab. The crack in the floor slab corresponded, roughly, to the line between the cut and fill areas of the site grading. The second soils engineer was again retained to investigate this occurrence. Again, he advised that the settlement would be of short-term duration, was largely complete, and that "no further cracks would appear or settlement occur." However, further settlement did occur. A total of 8 to 10 in of settlement had occurred when a third soil consultant was retained. He explored the site with conventional borings, but supplemented them by the use of large-diameter (30-in) auger holes, which permitted direct inspection of the soils. He found a layer of moderately organic SILT (alluvium) about 1½ to 3 ft deep under the strata of fill. He concluded that consolidation of this layer was the proximate cause of the observed settlement; that settlement would continue, but at a decreasing rate; that the building should be monitored; and that when the rate of settlement decreased sufficiently, that the building be "patched up." Judging by subsequent observations, the third soil consultant's conclusions appear to have been correct.

A suit was thereafter commenced by the owner against the contractor, the original soils engineer, and the consultant who performed the test borings. Quite noticeably the architect was not named by the owner as a defendant in this action. Damages of $1.25 million were asserted, alleging delay in completion of the store and repair costs to the building.

Because of certain procedural rules in the jurisdiction where the action was brought, the soils engineer was unable to join the architect as a party to this action. This prevented a direct claim against the architect for his alleged failure to have included vertical joints as a means of mitigating the detrimental effects of the settlement.

The testimony of the architect was then taken. He contended that the settlement cracks had resulted from a shifting or compacting of the alluvium and not from differential settlement. The architect clearly placed responsibility on the soils engineer for a failure to have advised the owner of the potential instability of the alluvium.

Experts were retained on behalf of the soils engineer to provide an analysis of this matter. The experts acknowledged that the original report failed to disclose the possibility of long-term settlement; nor was there any discussion of alluvium to enable modification of the foundation design to allow for possible settlement. However, the experts concluded that test borings taken by the consultant did reflect adequate subsurface materials and that similar results had been confirmed by the second expert retained by the owner.

Accordingly the experts retained on behalf of the soils engineer concluded that the report was a reasonable one and could be defended on its merits. They particularly pointed to the recommendation for the placement of vertical joints, which recommendation was not heeded by the architect.

In defense of the soils engineer, it also was stressed that the decision to continue construction after the cracks were originally noticed was that of the owner, the contractor, and the first soils engineer expert retained by the owner. It was this first expert who predicted, incorrectly, that no further settlement would occur.

During the course of the litigation, the owner failed to specify any particular act of negligence on the part of the boring consultant. Therefore, the owner's claim was dismissed. However, the first soils consultant was kept in the case by the cross claim of the soils engineer. The contractor continued to take the position that it had adhered to the plans and specifications in all particulars and bore no responsibility for the settlement cracks. The thrust of the owner's claim therefore fell upon the soils engineer.

In regard to damages, the owner produced evidence to show repair costs of $140,000. However, the owner was unable to establish any claim for delayed completion or loss of profits on the store. (The owner subsequently lost the building at a foreclosure sale.)

The action finally was settled when the owner agreed to accept $36,500 in total settlement of its claim. Of this amount, the soils engineer contributed $18,500, the contractor paid $15,000, and the boring consultant, who had been kept in the action by the cross claim of the soils engineer, paid $3000.

TECHNICAL ANALYSIS

The building in this case is a single-story structure constructed with exterior masonry walls supported on strip footings and interior columns supported on spread footings. Geologic sections are shown in Figure 26-1.

Prior to development, the site sloped from south to north, a total difference in elevation of about 80 ft (see Figure 26-1). In the 1930s, fill was placed over a portion of the site, forming a terrace at about elevation 1130 over the southern half, and sloping down to the north, as shown in Figure 26-1. Note the steep ridge shown in the profile, just south of the location where the main crack developed in the building. This fill was

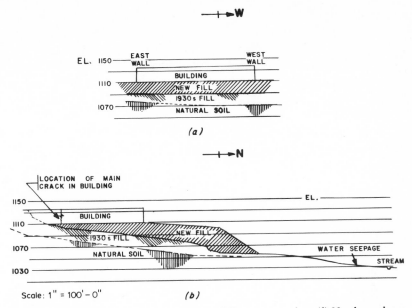

FIG. 26-1 Approximate site cross sections. (*a*) East-west section. (*b*) North-south section along west wall.

not an engineered fill, and in areas of the southern portion of the site tree stumps, roots, and other debris were found buried in the fill. However, in the area of observed settlement, exploration did not reveal such inclusions, but a clean material.

Site preparation for the building began in 1969. A bench was graded at elevation 1112 to accommodate the building. Cut operations in the southern area removed about 15 to 20 ft of old fill, natural soil, and rock. These soils were utilized to fill the northern portion. The thickness of the fill varied up to 40 ft. The side slopes of the fill pad were 1 vertical to 1½ horizontal. The new fill was well-compacted, roughly in accordance with the specified values of 98 percent and 92 percent of standard AASHTO, inside and outside the building, respectively.

During excavation for footings in the old fill in the southern portion of the site, quantities of organic materials and unsatisfactory soils were discovered. The area was undercut an average of about 13 ft to remove these materials and backfilled with compacted fill.

The history of the development of cracking in this building is contained in the Narrative section. The soil profile indicated in Figure 26-1 includes the following:

1. New fill, as described above.

2. Old fill, similar in properties to new fill, described as red-brown, silty,

fine to medium micacious sand of medium dense consistency. Cobbles and boulders of weathered mica schist were observed. No significant organic content was observed.

3. An alluvium, ranging in thickness from 1½ to 3 ft, representing the outwash from erosion of the old slopes (hill wash) on the site, was found in some borings. The alluvium is a moderately organic, sandy silt, in a stiff condition.

4. Residual soil derived from decomposition of the mica schist bedrock and described as a clayey, sandy SILT.

5. Bedrock.

A profile of the observed settlement is shown in Figure 26-2. Maximum settlement was about 8 to 10 in. Little or no lateral movement was observed. The settlement developed as a rotation about an axis along the line of the main crack in the building. Note that this crack is located close to the position of the ridge which traversed the site prior to the 1969 grading operations. Clearly, the filled area settled, whereas the cut area did not; the deeper the fill (the greater the added weight—ΔP), the greater the settlement. The following possible causes of the settlement were investigated:

1. *Settlement of New Fill* Compaction of the fill averaged about 96 to 97 percent, i.e., less than the 98 percent specified for material under the building, but not significantly so. Based on the borings, the fill was free of deleterious materials and was relatively dry (degree of saturation between 57 and 66 percent). It was concluded that the new fill was not a substantial contributor to the observed settlement.

2. *Settlement of Old Fill* Borings indicated that this fill (at least in the area of settlement) too was free of deleterious matter, was dry, and was very sandy. Any consolidation would occur rapidly, not over a period

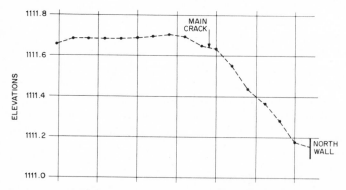

FIG. 26-2 Profile of observed settlement.

of a year or more, as had occurred. It was concluded that the old fill was not a substantial contributor to the observed settlement.

3. *Settlement of Natural Soils*   Consolidation tests indicated a consolidation of the natural soils at the site in excess of 0.05 in/in under an applied pressure increment of 4000 to 5000 lb/ft^2 (the weight of a 40-ft depth of new fill). Eight to ten inches of settlement would correspond to the compression of a 13- to 16-ft depth of the material—an evident possibility and the apparent cause of the observed movements.

4. *Subsurface Erosion*   This was due to the infusion of soil into buried utility lines. It was not considered probable that such an event could have caused movements of the type and area distribution which were observed.

5. *Lateral Displacement of the Hill Wash (Alluvium) Deposits*   Discounted as a probable cause.

Comments

Consolidation of Stiff, Fine-Grained Soils

The source of settlement in this case was traced (tentatively) to consolidation of the natural soils (hill wash and the residual soil). Both are described, essentially, as a sandy SILT in a stiff condition (N values were about 25 to 40). There tends to be some lack of understanding that even such stiff soils can consolidate substantially if the pressure increment (increase) is large compared to the original (preconsolidation) pressure (P_0). In this case, the consolidation tests suggest a value of P_0 of about 2000 to 3000 lb/ft^2. Placement of 30 to 40 ft of compacted fill added a pressure increment of 3500 to 4500 lb/ft^2, giving a value of $(P_0 + \Delta P)/P_0$ about 2.1 to 2.5.

Shallow Depth of Compressible Material

Note that the *third* soil consultant discovered the existence of a thin layer (1.5 to 3 ft) of alluvium (hill wash). This alluvium potentially was a more compressible material than any of the other strata encountered. The material was not detected in the previous explorations, which is understandable, since samples from borings are normally taken at 5-ft intervals. Note also that the third soil consultant, well aware that some problem existed, used 30-in diameter auger holes to explore the site and to make his discovery. Normally borings are made with 2½-in diameter casing, recovering 1⅜-in diameter samples. This problem of detecting thin, compressible strata is common. Most often it relates to the presence of 2 or 3 ft of meadow mat (peat) sandwiched between competent soils. The peat is highly compressible, with a 2- or 3-ft depth contributing nearly all of the anticipated settlement. The presence of silt-filled meanders or creeks in tidal or river flats is another example of this type of problem. There is no simple answer. The soils engineer must be a detective, reasoning from the history of the area and from the geology to determine what conditions to suspect.

Delayed Construction

This case illustrates a valuable point. If the project involves substantial fills, it pays (if practical) to delay construction of the buildings (or final grading and paving) until the fill has had an opportunity to "settle in." Some settlement plates to detect problem areas can be a worthwhile investment. Owners need to be educated to the advantages of two-step construction—fill as soon as possible and build later. Use the interim period to finalize plans, put in utilities and drainage, and put in the foundations, but pave or erect masonry after as much delay as is practical.

LEGAL ANALYSIS

In order to defend the soils engineer in this action, it was imperative that the defense be focused upon the warning given in the original report in regard to the need for vertical joints. Although the architect was not named as a party defendant to this action, it was argued that the owner was bound by the decision of its architect not to include such vertical joints in design. Here, the architect also did the structural design, and perhaps this, in itself, is a lesson not to be taken lightly.

The soils engineers' defense also included allegations that the damage caused by the settlement resulted from the decision of the first soils expert retained by the owner to continue construction without taking remedial measures. This type of defense is often characterized by attorneys as "trying the empty chair." This theory of defense contemplates pointing the finger of liability at a party who has not been named in the litigation, as a means of deflecting attention from a party who has been joined. The success of this defense varies in a direct ratio to the strength of the valid defenses which can be asserted on behalf of the defendant and the weakness of the position of the party who has not been joined.

It should be pointed out that in many states, the fact that an owner has sold the property in question prior to trial, and may even have made a profit, cannot be introduced into evidence by the defendants for the purpose of minimizing the owner's damage claim. The owner will be required to prove his or her damages, and the fact that the property may subsequently have appreciated in value can generally not be introduced as evidence by a party chargeable with having caused such damage.

case 27

Significant Factors

A. Failure to correctly evaluate magnitude of potential settlement

B. Failure to appreciate the adverse effects that the settlement would have on the usability of the building to the tenant

C. Failure to advise owner that more than the usual degree (1 or 2 in) of settlement would occur unless preventive measures were incorporated in the design

D. Failure of the architect-engineer to take control of the job; allowing the owner-developer to exert undue influence in the technical decisions required to produce a satisfactory, functional building

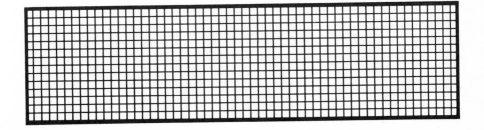

NARRATIVE

This project involved a settlement problem in the floor slab of a warehouse facility. The owner of the building retained the engineer to design this speculative industrial facility; his services were to include structural plans and specifications, but not to entail any supervisory or periodic inspection duties during construction. The engineer had been retained to perform his services in cooperation with the owner's own architectural and engineering department. The owner also acted on its own behalf as general contractor.

Various borings were made at the site prior to completion of the design. In addition to borings ordered by the owner, the engineer also commissioned his own set of borings. The borings commissioned by the owner appear to have disclosed the existence of highly erratic soil conditions sufficient to indicate that special attention would have to be given to the foundation design. The borings commissioned by the engineer resulted in a recommendation from the soils consultant for the use of a structurally supported floor. Such a floor would require an expensive process of driving treated timber piles past the layers of soft soil into soil capable of supporting the floor.

Although there was discussion between the engineer and the owner's architect concerning the need for a structurally supported floor, the architect instructed the engineer to use the less expensive method of a slab-on-grade floor. Plans for this type of design were completed by the engineer and approved by the professional staff of the owner.

The building was completed, following which serious settlement of the floor occurred, making it extremely difficult to fully utilize the warehouse. In a series of lawsuits by and between the owner and the tenants, a judgment in excess of $1 million for repair costs and loss of profits was rendered against the owner and in favor of the tenants.

A separate lawsuit was thereafter commenced by a successor to the owner (who had sustained the loss to the tenant) against the structural engineer and the contractor. Damages in excess of $2 million were alleged. The owner asserted that the engineer had failed to design the floor slab properly (inasmuch as the subsurface conditions were not capable of supporting a slab on grade).

In view of the finding in the action between the owner and tenants, which concluded that the owner was to bear responsibility for an amount in excess of $1 million, the major problem in defending the structural engineer revolved around the question whether or not the owner had sufficient information to determine, in its own right, that the use of a slab-on-grade foundation was improper for the facility in question. Clearly the structural engineer failed to follow the recommendations of the soils consultant with respect to the use of a structurally supported floor. An argument was raised that the engineer was never furnished with all documentation concerning the condition of the soils over the entire site; this information may have influenced a change in his recommendation in regard to the type of foundation employed. However, there was other evidence to show that the engineer was aware of the existence of highly compressible material (peat) in the soil at the work site. Consequently, a question of fact existed concerning the knowledge of the engineer with respect to the location and extent of the poor soil.

A related question bore upon whether the owner had received adequate warning from his engineer and/or the soils engineer to the effect that the slab-on-grade design was potentially unfit for this type of site and construction. After a substantial amount of discovery had been completed in the case, settlement negotiations were entered into by the parties. As a result, a settlement totaling $550,000 was reached, of which the contractor paid $50,000 and the structural engineer, $500,000.

TECHNICAL ANALYSIS

The building in this case is a single-story, warehouse type of structure constructed as a "speculative building," i.e., for rental, and without a specific tenant in mind. The building exhibited problems with the foundation, which consisted of timber pile support for the columns, and a slab on grade for the ground floor. The ground floor was raised to truck dock level, i.e., about $3'6''$ to $4'0''$.

As noted in the Narrative section, the slab on grade which formed the ground floor for the building settled. Settlement of the slab varied. A maximum of about $6\frac{1}{2}$ in of settlement had occurred when the tenant instituted its suit against the owner. Remedial measures consisted of removing the floor slab in the area of greatest settlement and replacing it with a pile-supported, structural slab. In more stable areas, a topping or leveling course was added to restore the floor level.

On the surface, this is just another case of an engineer unwittingly building in an area underlain by organic soils, without proper appreciation for the magnitude of the settlements which would occur and, therefore, without taking necessary preventive measures. Though other, similar failures have been described in other cases in this text, this case contains one new element. The engineer involved with the design was a prominent foundations expert. Ignorance of the significance of the presence of the organic soils is most unlikely. What happened? The answers to that question carry several lessons for the design engineer, the owner, and for prospective tenants.

1. The plan area of the building is about 65,000 ft². The initial exploration program consisted of five borings, i.e., one for every 13,000 ft² of build-

ing area. Based on these five borings, the geologic section shown in Figure 27-1 may be inferred. (Actual logs of the applicable borings are included for reference, as Figures 27-3 to 27-5.) Note that one boring per 13,000 ft^2 of building area is woefully inadequate by most modern standards (which require one boring for every 2500 ft^2 of building area,[1] i.e., about 5 times as many borings as were taken in this initial program). To his credit, the designer noticed the indication of "peat" in this initial set of borings and requested seven additional borings, for a total of twelve, i.e., still less than half the normal requirement. However, the developer reduced this request from seven to five.[2] Whether or not the designer protested this reduction and, if he did, how strongly he protested is not known. The end result was that in a sensitive situation, about 40 percent of the normal requirement for borings were made. Why should a building code specify the number of borings? Why not performance criteria such as, Enough borings to suit the situation? One reason, sad to say, is to protect engineers from the pressures which may be exerted upon them by their clients.

2. During the course of investigation of the cause of the settlement of the floor slab, more borings were made. The N-S geologic section inferred from these borings is shown in Figure 27-2. It should be compared with Figure 27-1. The material described in the original borings as CLAY, with N values of 10 to 15 (3 to 8 in boring no. 41) now appears as PEAT and organic silty CLAY, with N values of 1 to 2. Incredible! It is as though the two sets of borings were made at different sites. It

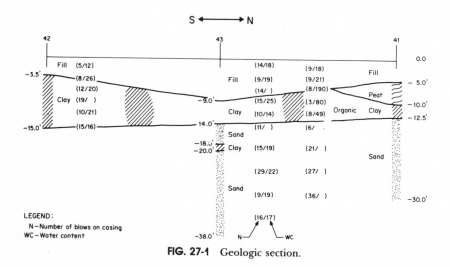

FIG. 27-1   Geologic section.

[1] See Building Officials Congress of America (BOCA) Code for example.
[2] One of these additional borings also showed the presence of peat.

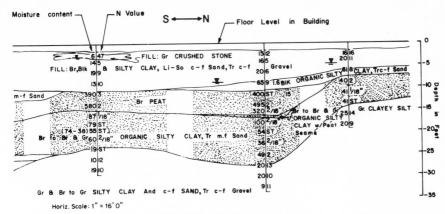

FIG. 27-2 Geologic section.

is because of occurrences such as this that on-site inspection of borings by the party responsible for the foundation design is advisable and, in some cases (New York City Building Code, for example), is a legal requirement.

3. Depositions in this case suggest that the engineer knew that settlements due to consolidation of the peat would occur and the engineer claimed that he orally advised the owner-developer to excavate such peat within the building area and to backfill with select material before erecting the building. The owner-developer denied receiving this advice, and the judge in the case found that this instruction, if given, was not effectively communicated to the owner-developer—there being no written record of such a communication. The lesson? An old adage: If it's important, put it in writing.

4. In any event, the engineer claimed anticipation of 2 to 4 in of settlement and that a floating floor slab which would settle by this amount, with the periodic use of leveling courses to restore grade, would produce a perfectly usable warehouse. For a bulk storage warehouse, such as a transit shed, this is true, and the author has successfully designed several such buildings to such criteria. But in this case two things happened. First, analysis indicated that 2 to 2½ ft (not in) of settlement could be expected. Second, the tenant did not use the premises as a warehouse, but for the manufacture of corrugated boxes. Presses and other machinery and the storage of rolled paper were sensitive elements affected by even a few inches of settlement.

5. That the premises were leased by the owner-developer to a tenant whose needs were incompatible with the construction and without the tenant being made aware of the potential incompatibility was one of the issues in the case. We do not propose to comment on the owner's responsibili-

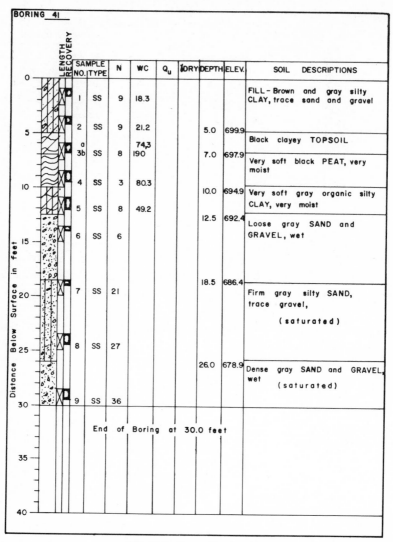

FIG. 27-3   Log of boring 41.

ties in the matter. That is covered under Legal Analysis. We only wish to point out that the tenant can do much in the way of self-protection. Most plans contain a set of "General Notes" which list the structural design criteria and often a table of "Design Live Loads." It is a simple matter to check if the criteria described suit the tenant's needs. Most plans also contain a set of boring logs and a boring location plan. It is a matter of a few hours work for an experienced foundation engineer

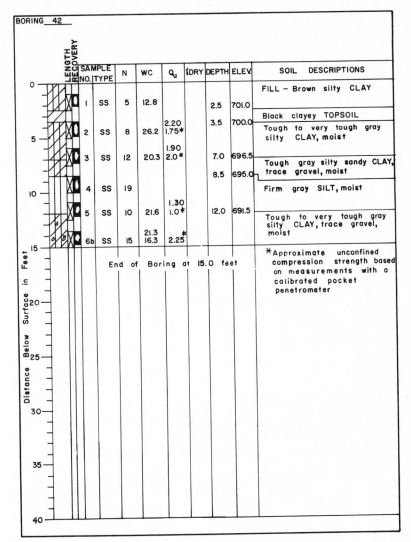

FIG. 27-4 Log of boring 42.

to review this data (and the foundation plans) for purposes of detecting manifest or gross deficiencies.

6. Much attention was paid in the discovery proceedings in this case to establishing the design live load used in detailing the slab on grade. Numbers from 100 lb/ft² to 1000 lb/ft² appear in correspondence and testimony. Terms such as "warehouse," "light manufacturing," and "heavy manufacturing (industrial)" are used. Overlooked is that toler-

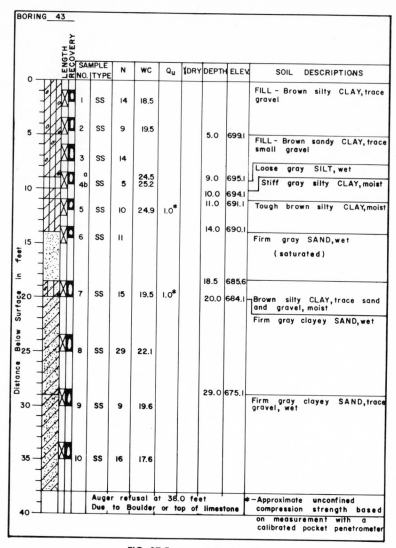

FIG. 27-5 Log of boring 43.

ance to settlement of a slab on grade is not related to the load per se, but to the use of the premises. Also overlooked was that the weight of the fill necessary to raise the floor level to dock height was sufficient to cause 6 to 8 in of settlement, without any contents in the building.

7. Refer to Figure 27-2. Removal of the soft, compressible soil would have required up to a 30-ft depth of excavation. A 15-ft depth of excavation would have been required to remove just the peat, leaving the organic clay. Consolidation test data indicated that, of the 2 to 2½ ft of potential

settlement, half, or more, would be due to consolidation of the organic clay. Therefore, removal of the peat, even if mandated by the engineer, never was an adequate solution in itself. Why did the engineer not realize the magnitude of the settlement potential represented by the organic clay? The high N values indicated in Figure 27-1 may have been the reason (later borings did not corroborate these high N values). In any event, the admonition of Case 26 applies to this case, as well. If the soil profile shows a substantive depth of silt or clay soil, consolidation tests are likely to be a wise investment.

8. Finally, there was physical evidence at the site of the existence of a swamp and that the building area was a fill placed over the old swamp. Testimony indicates that neither the engineer, nor anyone else responsible for the foundation design, ever visited the site of the work. A visit and inquiries to appraise the geologic and historic background of a proposed building site is always advisable when working in a new area.

LEGAL ANALYSIS

The exposure which the engineer faced in this matter concerned whether he was negligent and/or breached his contract by failing to recommend the design concept that would have been more expensive than the slab-on-grade design ultimately utilized. On behalf of the engineer, it was argued that he had limited professional duties on the project and that his design was subject at all times to the review and approval of the architectural and engineering departments of the owner. A provision to this effect was specifically contained within the contract for services between the owner and the engineer.

It should be noted that the project discussed herein was constructed for speculative investment purposes, and hence the owners were quite clear in their instructions that they were always seeking the least expensive methods of construction. Of note is the fact that the final drawings for the building were signed and sealed by the architect employed by the owner, evidencing that the plan had been prepared and reviewed under his supervision.

From the standpoint of the owner, it was asserted that while it knew that problems with the soil existed, adequate steps had been taken on its part to avoid any problems. These steps included directions to the engineer to secure additional soil borings. In a decision handed down by the court with regard to the action between the owner and the tenants, a finding was issued to the effect that the owner's architect was aware that without a structurally supported floor the utility of the building would be limited. However, doubt was cast upon the contention of the engineer that he had fully communicated these warnings to the architect.

Clearly there was a close relationship between the engineer and the architect. All decisions relating to the design were closely coordinated between the staffs of both offices. Certainly, the architect was deeply involved in all design decisions and appeared to have been fully aware of the problems with the soil at the site.

Problems with defending a case of this nature center around the reasonable belief of counsel that a jury hearing this muddled controversy would hold each party equally responsible for all damages proven at time of trial. Far worse, it is always possible that witnesses for one side will subjectively appear more credible to a jury, thereby resulting in a finding totally in favor of one side or the other.

It is upon this basis that counsel often are required to make experienced judgments in advising their clients regarding the expediency of proceeding to trial or attempting to negotiate a settlement on favorable terms. Many factors go into comprising such a recommendation, including the cost of continued litigation, the potential maximum exposure faced by the client, the time which will be spent by the client in preparing for and attending a trial, and the financial impact that an adverse decision would have upon the client.

The settlement achieved in this matter represented less than one-half of the damages which had been sustained by the owner in the earlier action brought by the tenants. Clearly damage had been incurred. The findings in the earlier court decision between the owners and the tenants had not clarified the issues sufficiently to warrant a belief that the engineer would ultimately prevail in this action. Accordingly, a negotiated settlement of less than 50 percent of the damages incurred represented a reasonable compromise of this matter.

case
28

TYPE OF FACILITY　Pavement for plaza

TYPE OF PROBLEM　Cracking and heaving of terrazzo topping

Significant Factors　**A.**　Improper location and detailing of joints

B.　Lack of coordination between structural engineer (who detailed pavement base slab) and architect (who detailed topping)

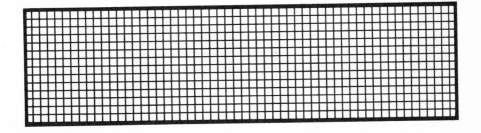

NARRATIVE

The project in this matter involves a large convention center containing a free-form plaza area which included an ice skating rink in its southerly portion. The geographic locale of the project placed it in an area subject to severe, snowy winters and hot summers.

The northerly section of the plaza was constructed by placing a 6-in concrete slab reinforced with welded wire mesh over a layer of broken stone. A 1-in concrete topping was then placed between a grid of brass stripping (9- by 3-ft module), and was bonded to the base slab. The southerly section was comprised of a base slab of 9-in concrete. This slab was also placed over a layer of broken stone and topped by a 1-in terrazzo finish with no brass strippings. As this southerly area was to be used for the ice skating rink, cooling pipes were cast within the top 4 in of the base slab.

Problems developed shortly after completion of the project when the terrazzo in the southerly section of the plaza cracked. The terrazzo topping separated from the base slab and heaved. Similar heaving occurred in the northerly section. Damage was so severe that the entire topping in both areas had to be removed. Separate investigations were conducted by the owner, the contractor, the architect, and the structural engineer, all of whom were involved in design and construction of the plaza. Experts were ultimately retained on behalf of all the involved parties.

It quickly became clear that there had been a number of separate factors which may have contributed, in whole or in part, to the cracking. As no party agreed to take primary responsibility for the problem, the owner requested the contractor to remove and replace the topping. The contractor refused to do so unless he was given a separate written contract which was ultimately agreed to by the owner.

Prior to removal of the topping, all parties made visits to the site, took photographs and samples, and performed other tests. It was fully anticipated that this case would be litigated.

The contractor denied any liability for the cracking. As for causation, the contractor and his experts pointed to the fact that the design called for construction joints in the base slabs with no matching joints in the terrazzo topping. There was also recognition that the topping and the base slab lacked adequate bonding at several locations, thereby eliminating resistance of the topping slabs to buckling.

From the standpoint of the architect and the engineer, the contractor's liability was predicated upon a lack of proper preparation of the base slab by failing to attain adequate bonding for the topping. This involved an apparent failure by the contractor to scarify the surface of the slab before placement of the topping. Additionally, the design team pointed to an apparent lack of proper curing of the slab surface to obtain proper hardness and durability, the use of unapproved polyester material as a setting bed for the brass divider strips which were placed in the topping at the north end, and a failure to submit shop drawings of the slab construction joint for the architect's approval. According to the architect, without such a review, it was impossible to coordinate the location of the slab joints with the topping brass divider strips, thereby effecting cracks in the topping.

In regard to the potential liability of the architect and engineer, it became critical to determine the nature of the services each provided with regard to the plaza area. The architect, of course, was responsible for the ultimate design and coordination of plans. However, the engineer did have a written agreement calling for it to provide inspection of construction relating to structural items and to furnish a full-time resident engineer during this construction phase. The contract also stated that during this inspection, "the engineer shall endeavor to guard the owner and architect against defects and deficiencies in the structural work of the contractor." It should also be noted that the owner retained a full-time inspection team and construction manager to insure compliance with plans and specifications.

Thereafter, the contractor filed a claim for $1.6 million against the owner for alleged delays and interference. The contractor attributed a total of $162,879 as its out-of-pocket costs attendant to the removal and replacement of the topping. Despite this claim, meetings were held by and between the various parties in an effort to resolve the dispute. The owner did request assistance from the design team for purposes of opposing the delay claim from the contractor. This information was provided by the architect and engineer.

Ultimately, an action was commenced by the owner against the contractor, the construction manager, and the architect. The architect joined the structural engineer as a third-party defendant. The total replacement cost which was incurred on behalf of the owner was $217,000.

Prior to the commencement of this action, counsel for the various defendants had met on numerous occasions to try and resolve their differences. These meetings continued after the filing of the lawsuit over a period of several months. During the course of these discussions, an understanding was reached whereby each party recognized that it was in the best interest of all defendants to attempt to have the owner decrease its damages to a point where contributions from the defendants could be secured to make a reasonable offer of settlement. This was subsequently achieved and settlement was effected in the sum of $150,000. The owner agreed to discount its total out-of-pocket loss substantially, the architect and engineer agreed to contribute $58,500 each, the contractor agreed to pay $26,000, and a contribution of $7000 was secured from the construction manager.

TECHNICAL ANALYSIS

The sympathetic cracking of the topping, corresponding to the movements in the base slab, as described in the Narrative, is not likely to be a surprise to any experienced engineer. It is a well-known phenomenon. Why was it not anticipated in this case? Two possible explanations derive from reading the case files.

1. The topping was an architectural treatment and detailed by the architect, who may not have been as familiar with the potential problem as an engineer would be.
2. Something may have been lost in the evolution of the detailing of the topping. The initial specification called for a ½-in topping, shake-troweled into the still plastic concrete. This was the most logical procedure to assure good performance under the local climatic conditions. Later, the specification was changed (reason unknown) to call for a 1-in topping, between brass divider strips, with the divider strips set in the plastic concrete. This was not practical to do from a construction standpoint. The final specification called for a 1-in topping, between brass dividers, set on the hardened concrete—and this is what was built. However, such a detail makes it difficult to assure a good bond between the topping and the base slab. The reasons are (1) the divider strips are an obstruction to the surface preparation procedures, (2) a good bond between a slab and a thin overlay is difficult to obtain in any circumstance and the divider strips made this more difficult, and (3) the divider strips did not function as movement joints.

Why the design proceeded from a detail which was more proper for the circumstance to one which was less assured is not known. Rather, the basic elements required to produce a good floor topping and some cautionary words are summarized below. For greater detail, refer to a series of articles in *Concrete Construction,* published between 1969 and 1973, and to ACI 302 "Recommended Practice for Concrete Floor and Slab Construction."

1. Unless shrinkage-compensating cement is used, or posttensioning is used sympathetic cracking in the topping (with cracks and movement joints in the base slab) is inevitable. Movement joints in the topping should be stacked over the corresponding joints in the base slab.
2. A good reliable bond between an overlay (bonded topping) and the base slab is difficult to achieve. For bonded toppings, a mechanical keying, in the form of surface roughness, is essential. If not provided in the slab finish, mechanical hacking is required. The bonding surface must be absolutely clean. Scrubbing, hosing, even an acid etch are re-

quired. The surface must be free of all excess water—but not "dry." The difference in temperature between the base slab and the topping mix should not exceed 10° F. A bonding agent (grout, epoxy, or other material), well worked into the slab is necessary.

3. Finishing is a key element in the performance of integral toppings. The topping material should normally be pounded into the surface with a float and worked only until barely covered with the cement paste. This should be followed by steel troweling to compact the surface. Some sources recommend doing the steel troweling twice to assure compaction. Power finishing is desirable.

4. Since sympathetic cracking will occur, crack control in the topping is directly related to the provisions for crack control in the base slab. Figure 28-1 illustrates some of the basic principles.

5. Joints must be sealed to prevent the intrusion of dirt, which prevents closing movement.

6. Panels between control joints should be as nearly square as is practical.

7. Shrinkage control—in the form of low slump, water-reducing admixtures, controlled placement sequence, and proper curing are essential.

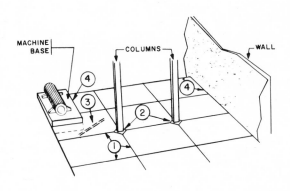

MACHINE BASE   COLUMNS   WALL

① CONTROL JOINTS – 15' TO 25' GRID.

② ISOLATION JOINTS AT COLUMNS.

③ ADDITIONAL REINFORCEMENT AT REENTRANT CORNERS.

④ ISOLATION JOINTS AT WALL AND OTHER POINTS OF RESTRAINT.

FIG. 28-1 Base slab.

LEGAL ANALYSIS

The settlement achieved in this matter reflected the mutual respect between both the parties and their attorneys and the sophisticated understanding of the problem evidenced by all concerned. Throughout the course of the numerous meetings that were held, each party and its counsel discussed the technical problems each party had in defending its position and attempted to deflect responsibility away from itself. Detailed discussions regarding damages, repair costs, and apportionment of liability enabled all concerned to reach a mutually agreeable settlement proposal.

At the outset of these negotiations, the owner took the position that the design team was 90 percent responsible for the problem, basing this judgment upon an expert report it had commissioned. Needless to say, the contractor took a similar position, and for a protracted period of time was adamant against making any contribution.

Between the architect and the engineer there were extensive discussions designed to defend each party's input. It was quickly realized that neither the architect nor the engineer could totally justify the lack of coordination which was evident in the design of the topping and base slab. The experts clearly indicated that the architect would be primarily responsible for the architectural aesthetics of the topping; they likewise attributed all ramifications of the design for the base slab to the structural engineer. The failure of each to coordinate its design with the other probably represented a joint liability. Hence, the design team decided to join forces in an attempt to secure contributions from the contractor and the construction manager in order to effect a settlement.

In this case, the only active litigation involving the parties was the filing of the summons and complaint served upon the contractor, construction manager, and architect, followed by the third-party complaint against the structural engineer. Counsel for all parties agreed that proceeding through discovery, the review of the documentation promulgated by the project, as well as taking standard depositions of all parties, would not adduce any substantial information not already known to the parties. Hence, there was general agreement that this phase of the litigation was unnecessary. This agreement probably saved as much as $100,000 in litigation costs.

It should be mentioned that not all cases lend themselves to this type of agreement. Quite often one party is disadvantaged by not having access to all pertinent documents and must proceed through discovery to secure this information. Moreover, each of the counsel involved were attorneys sophisticated in construction matters and there was little expenditure of time needed to educate the attorneys as to the problems at hand.

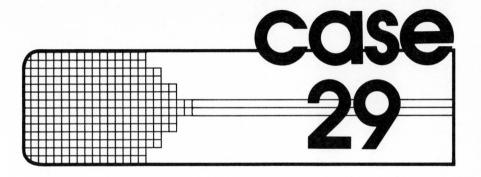

case 29

TYPE OF FACILITY Prefabricated metal building

TYPE OF PROBLEM Leakage through metal roof deck

Significant Factors
A. Faulty workmanship
B. Design deficiencies
    1. Inadequate stiffness in stiffened lip
    2. Inadequate splices in purlins
    3. Inadequate roof slope

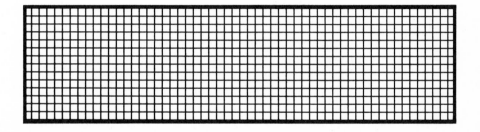

NARRATIVE

The building involved in this action was a warehouse and adjacent facility for the manufacture of glass bottles. Following a severe snow storm and sudden thaw, extensive leaks developed in the warehouse facility. As a result of those leaks, the owner engaged two independent consulting engineering firms to recommend solutions and to conduct evaluations of the design and construction of the warehouse.

The building was a prefabricated steel structure designed by an engineering and fabrication firm and constructed at its plant in accordance with a written contract executed by the owner. The prefabricated building was then purchased by a general contractor who arranged, pursuant to a subcontract, for erection of the building at its site.

The severity of the leakage was such that practically all seams of the building exhibited water penetration.

As a result of various engineering studies which were performed in order to remedy the roof situation, other design-construction problems were brought to the attention of the parties. The most significant design problem was that the calculated lateral deflection (or sway) under the design wind load was 11.6 in. The other design problem encountered involved overstressing of approximately 50 percent in some braces and as much as 52 percent on one exterior column. In addition, evidence existed to show that the roof would fail to meet the design load test weight of 40 lb/ft² since the purlin lap, as detailed, was insufficient owing to a miscalculation in the engineer's design.

From a construction standpoint, inspection indicated that fasteners at the roof level did not conform to the specifications of the engineer. In some instances, only half the required fasteners were in place.

Experts for the engineer intimated that the roof leaks were the result of a ponding condition created by the accumulation of water when snow on the roof acted as a dam, and, ultimately, testing at the site by these experts for the engineer confirmed that the roof leak was caused by such a ponding condition. The engineer faced substantial exposure for the leakage situation.

Negotiations between the owner, engineer, contractor, and erector were commenced following the completion of the inspection and testing program. An initial demand was made by the owner for a settlement somewhat in excess of $600,000. However, there were indications that a settlement of substantially less than this sum would be accepted. No damage to the interior of the facility or stoppage of production was asserted and this acted to substantially reduce the total amount of claims asserted by the owner. From the standpoint of the engineer, an analysis of the potential damages indicated that an ultimate judgment, were the case to proceed to trial, could be awarded in the area of $750,000 to $1 million. Hence, settlement negotiations designed to secure the resolution of this claim were deemed to be in the best interest of the engineer.

Problems were encountered during the negotiations when the erector and the contractor both refused to become actively involved in the settlement discussions. The erector advised that it had no insurance coverage for the damages involved, and had financial resources to contribute little more than a nominal amount toward a settlement. The contractor likewise advised that it lacked insurance coverage. However, the owner had retained approximately $60,000 in fees due the general contractor for the erection of the facility. The general contractor subsequently agreed to contribute this amount in return for a full release for its services on the project.

Settlement was thereafter effected for a total sum of $440,000, of which the engineer contributed $415,000 and the steel erector contributed $25,000. This amount was exclusive of the $60,000 which had been retained by the owner for fees due and owing to the general contractor.

TECHNICAL ANALYSIS

A diagrammatic section through the building in this case is shown in Figure 29-1. The source of leakage was traced by the use of ponding tests on the roof and was found to originate in gapping of the side and end laps of the roof decking, as shown in Figure 29-2. The proximate causes of the gapping were determined to be as follows:

1. End Laps
    a. Excessive flexibility of the purlins: The purlins were designed as continuous beams. Periodic splices were required. Lap splices were used. The calculated, required lap length was 4'8". The lap length as installed was only 2'4". As a result, the connectors in the splice were overstressed by about 70 percent, and the splices rotated (see Figure 29-2a). Deflection and end rotation of the purlins were markedly noticeable during the ponding tests.

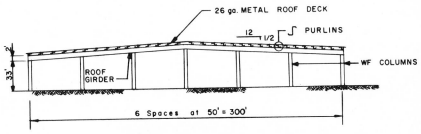

FIG. 29-1 Typical section of building.

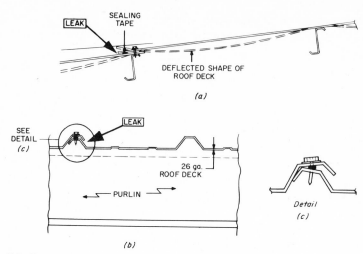

FIG. 29-2 (*a*) Leakage through end laps. (*b*) Leakage through side laps. (*c*) Gapping due to spreading of sheets of roof deck.

b. Inadequate roof pitch: Specified roof pitch was ½ in. in 12 in and the roof, overall, was so built. However, the roof slope was uneven so that the slope in some areas was less than ½ in. in 12 in, facilitating penetration of water under the action of the wind or, where accumulations of snow occurred, the ponding which resulted. Also, a backup of water (ponding) occurred when the gutters became clogged with ice and snow.

c. The end lap was not installed as per the design (see Figure 29-4). The length of lap, as installed, was less than as indicated in the design and, in some locations, *the sealant was installed in the wrong location.*

d. The proper number of screws had not been installed in the end laps.

2. Side Laps (Leakage through side laps was substantially less than through end laps)

   a. Side laps showed lack of caulking.

   b. Deflection of the roof sheets, under load (which was highly evident) caused the side laps to spread. They did not return to their original configuration after removal of the load.

   c. Some of the side laps were poorly nested.

Other issues (which will be of interest as a checklist of precautions) raised by the investigations incident to the subject case were as follows:

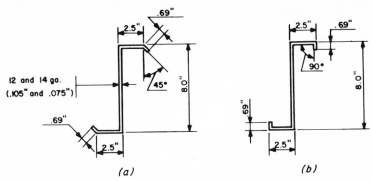

FIG. 29-3 Detail of purlins. (*a*) As built. (*b*) Required to meet requirements of AISI Code.

1. The stiffened lip on the compression flange of the purlins (see Figure 29-3*a*) did not conform to the requirements of the American Iron and Steel Institute (AISI) Code. Note the minor modification which would have been required to satisfy the AISI Code.

2. Roof had not been cleaned of metal filings from self-tapping screws. These filings interfered with the close nesting of the sheets.

3. Snow buildup at eaves and at the adjacent building was not considered in the design.

4. No computations were made for a number of the connections in the building framing. This is a regrettable practice.

5. No investigation was made in the design of uplift (suction) forces.

   Correction of the condition consisted of installing new metal decking (and insulation, where it had existed). New and additional bolts were installed in purlin splices and existing bolts were tightened. Although not related to the leakage problem, but in order to correct deficiencies discov-

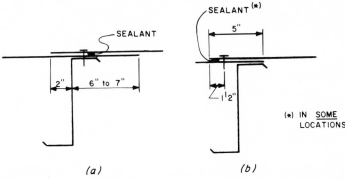

FIG. 29-4 Detail of end laps of roof sheets. (*a*) As designed. (*b*) As built.

ered during the investigations which followed the incident, miscellaneous, minor revisions and additions were made to the basic framing and bracing system.

LEGAL ANALYSIS

The function of an attorney in a case where experts have concluded that design defects were the cause of a substantial amount of damage, is to guide the client to a resolution of the case in a cost-efficient and pragmatic fashion. There is little to be gained by putting on a strong show of bravado or legal histrionics. An attorney will easily lose credibility with the other parties if he attempts to deny a fact readily determinable upon investigation. Moreover, in a situation where the liability of the contractor and subcontractors is limited, it is incumbent upon the attorney for the design professional to take the lead in determining the most efficient manner of resolving the dispute.

The overriding factor in this case was a determination in regard to the damages which would be recovered by the owner for the roofing problems encountered on this project. Any analysis of roof damage such as is detailed above, is highly technical and laborious. There is nothing glamorous in poring over technical materials, expert reports, and construction estimates as a means of determining prospective costs which can be anticipated for remedial work necessary to restore a failure of this type. Nevertheless, such work must be undertaken and the success of negotiations with the owner will hinge upon the means by which damages can be reduced to the barest minimum.

Here, in addition to working out the damages, counsel was also required to explore the possibility of securing a release for the engineer for any possible personal injury claims to workers which may have resulted from the roof damage. Working with counsel for the owner, the records of the operation were reviewed and it was determined that no such injuries had been sustained. Full releases were secured from the owner with the exception of potential bodily injury claims involved in prior construction or erection of the facility. Moreover, it was incumbent upon counsel to draft an appropriate release which would preclude future payments for design defects which were latent in nature and hence not discernible by virtue of the inspections undertaken by the owner's experts to that time.

In essence, it is incumbent upon the attorney for a design professional to consider not only the extent of the damage incurred as a result of the failure, but to make a concerted effort to limit any exposure which the design professional might face as a result of past or future incidents related to the one at hand.

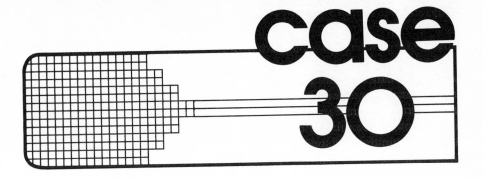

case
30

Grain silos

Excessive stresses in ring girder
supporting hopper bottoms

Significant Factors   Failure to evaluate torsional stresses

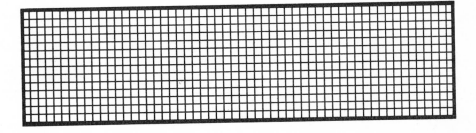

NARRATIVE

The following case represents an example of the need for architects and engineers to ensure that careful compliance with their contractual obligations is carried out during the course of design and construction. Of course, it is essential that design professionals fully understand the terms of their agreement and not rely upon mere "boiler-plate" language which they have historically included in their contracts.

The project involved a bulk cargo handling facility. Specifically, problems occurred with respect to the design and construction of the silos and grain elevators. The owner had retained a general contractor who, in turn, subcontracted construction of the silos to another contractor. The owner had also retained an engineer to design the facility and to supervise construction to assure conformance with the design.

The engineer's specifications for the silos and hopper bottoms required that the subcontractor prepare the design in accordance with the AISC Manual. The specifications also required that the design of these facilities be signed by a registered structural engineer who would have his drawings and calculations submitted to the owner's engineer for review to ensure compliance with the overall design concept of the facility as well as to ensure compliance with the specifications prepared by the engineer.

The subcontractor thereafter did retain a structural engineering consultant to design the silos and hopper bottoms. This consultant prepared the design and sent it to the engineer, who requested certain changes. These changes were made and the drawings resubmitted to the engineer. The design provided for the steel cones of the hopper bottoms to be connected by welding to a T-shaped ring girder, with the point of attachment at the bottom of the girder which was, in turn, supported by steel I beams (see Figures 30-1 and 30-2).

Following the submission of this original design to the engineer, the subcontractor learned that he was unable to obtain a steel fabricator who could roll the girder in the T-section form as initially designed. The consultant thus was asked to redesign the girder so that it could be fabricated from rectangular sheets of steel plate which would be welded together to form a box section.

In redesigning the system from a T girder to a box girder, the consultant altered the point of attachment of the cone by providing for the cone to be attached and welded at the top of the girder, rather than at the bottom. This redesign was sent by the consultant to the engineer. The engineer's staff checked the new design and noted that the point of attachment of the cone at the top of the girder was inconsistent with the engineering calculations previously received and that there were also certain other irregularities in the revised design. They first notified the consultant that certain features were considered objectionable and asked that they be corrected.

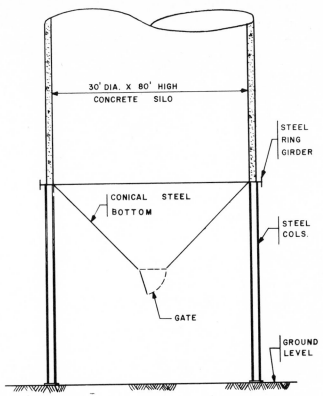

FIG. 30-1 Schematic of conical steel hopper bottom suspended from ring girder.

This was done and new design drawings, together with the engineering calculations supporting them, were prepared and transmitted once again to the engineer.

In preparing the new calculations requested by the engineer, the subcontractor's consultant ignored the torsional stresses in the box girder. He did this out of a belief that the attachment of the cone was at the bottom of the girder, as he had originally designed. However, the actual attachment of the cone at the top of the box girder induced torsional stresses greatly in excess of the permitted limits set forth in the AISC Manual, and these stresses were of such nature as to make the system, as designed, unsafe.

The engineer checked and reviewed this revised submission and, apparently, also missed the problem of torsional stresses, and therefore found the revised design to be satisfactory. In what later turned out to be a point of contention at trial, it was discovered that prior to receiving the revised engineering calculations and drawings, the engineer had filed the original design as stamped and sealed by the consultant for approval by

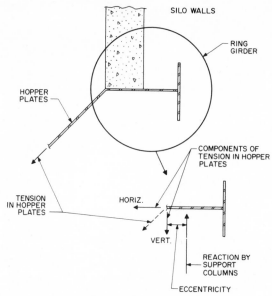

FIG. 30-2   T-shaped ring girder subject to forces.

the local building department. This was done in support of an application for a building permit to construct the facility. However, the revised drawings and calculations prepared by the consultant and subsequently approved by the engineer were never filed with the building department.

During construction, the misdesign was discovered and work on the hopper bottoms was suspended. The engineers investigated and determined that the box ring girders and cones of the hopper bottoms required correction. This was done. Additional costs and delays in construction resulted.

In subsequent litigation, it was alleged by the owner and the contractor that the cause of the misdesign was the responsibility of the consultant and the engineer jointly, and that they were, accordingly, responsible for the delay damages and increased costs of construction which resulted.

A substantial lawsuit was commenced by the contractor against the owner, the engineer, the concrete subcontractor, and its consultant. A trial lasting several months resulted in testimony ranging over every detail of the design configuration and ultimate revisions and approvals concerning the hopper bottoms. Although not here germane, there were numerous other claims relating to delays, welding problems, approval of progress payments, and attendant damages for these items.

At time of trial, testimony was taken from the consultant and representatives of the engineer. The consultant admitted that he had not calculated the effect of putting the connection of the ring girder and hopper cone at the top section of the ring girder. He stated that when he later recalculated the forces acting upon the connection he determined that it was overstressed and that this was a simple engineering mistake. He acknowledged that the overstress occurred due to the torsional stresses, that he had not known about such stresses previously, and that he had relied upon a technical treatise which referred to stresses in a hopper bottom, but which likewise had failed to consider torsional stresses.

The consultant also acknowledged that he had been notified by the engineer that his calculations, which showed a T-section ring girder, and his drawings, which showed a box section ring girder, were inconsistent, and that he was requested to provide additional calculations. He acknowledged having prepared these calculations on the box girder and forwarding them to the engineer. He testified that these calculations assumed that there would be an attachment of the cone to the ring girder at the lower portion of the ring girder. He also stated that he made no calculations concerning attachment of the cone to the ring girder at the upper portion of the ring girder and that when the problem was detected, he did make these calculations and thereupon realized that the ring girder was overstressed.

The testimony of the engineer's representatives cast a different light upon the manner in which the drawings and calculations were prepared and approved. Essentially, the engineers testified that they made only a cursory review of the consultant's calculations, and were not required to do a thorough recalculation by their contractual agreements with the owner. The chief design engineer testified that when he received the drawings and calculations, he briefly reviewed them and turned them over to a junior engineer in his office. Significantly, the junior engineer was not a registered engineer, having only recently received his engineering degree. The senior design engineer testified that, based on his previous dealings with his subordinate, the latter was competent to make a review of the drawings and calculations of the consultant.

On behalf of the engineer, it was argued that their contract called upon them to review all shop drawings only to determine whether they conformed to the general design concept of the project. It was argued that shop drawings were defined as those drawings which come from a subcontractor, and which are not deemed to be samples. Therefore, the consultant's submissions could be considered as shop drawings. However, the approval of the engineer with respect to the drawings and calculations of the consultant was made pursuant to the specification which was incorporated in the contract providing for the subcontractor to have the tanks and hopper bottoms designed by a registered structural engineer and that such drawings and calculations would be sufficiently detailed to show the work to be done. These specifications were to be approved by the engineer. Accordingly, a substantial question of fact and law existed with regard to the nature and extent of the approval given by the engineer for these drawings and calculations. Clearly, if it were determined at the trial that the engineer was required only to review the design and calculations for the hopper bottoms to ensure compliance with the overall design, any error in the design would rest upon the consultant. If, however, it was determined that approval by the engineer was to be undertaken in accordance with specification incorporated in the contract calling for detailed review, a joint and several liability would rest upon both the engineer and consultant for the design error.

After several months of trial, a determination was handed down by the court awarding damages against the consultant and the engineer amounting to $146,000, inclusive of interest and attorney's fees. Appeals were filed, alleging numerous errors in the findings of fact and conclusions of law filed by the court. The matter was ultimately settled for the sum of $132,000 on behalf of the two responsible parties.

TECHNICAL ANALYSIS

The subject structures were twelve elevated storage silos (for grain). Construction consisted of circular silos (of concrete) supported by a steel ring girder, carried on steel columns. A conical steel hopper bottom was suspended from the ring girder. The construction is shown schematically in Figure, 30-1.

In such a structure, the ring girder is subject to forces, as shown in Figure 30-2. The horizontal component of the tension in the hopper plates is resisted by the "ring" action of the ring girder. The vertical component is delivered to the support columns via bending in the ring girder. The section of the ring girder assumed to resist this bending might be assumed to include only the T material shown, or to include some interaction by the hopper plates or by the silo walls as the designer might consider appropriate. The eccentricity between the vertical component of the reaction in the hopper plates and the reaction of the support columns causes a torsion in the ring girder. The amount of eccentricity and the magnitude of the torsion depend on the details of the ring girder and the details of the joint assembly (hopper plates and column locations).

Initially, the designer detailed a T-shaped ring girder, on the general scheme of that shown in Figure 30-2. Difficulty reputedly was encountered in finding someone to fabricate such a section, and the designer changed to a box shape for the ring girder, roughly in the configuration shown in Figure 30-3a.

Investigation of the case revealed that the designer had intended that the hopper plates be connected to the ring girder at the bottom of the girder, presumably as shown by (1) in Figure 30-3a. With such a location, the resultant of the applied loads acts, more or less, through the centroidal

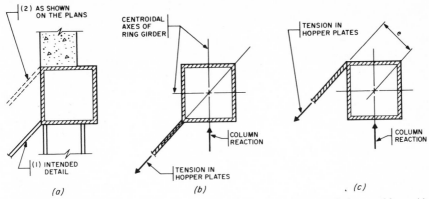

FIG. 30-3 (*a*) Ring girder, as designed. (*b*) Forces acting on intended assemblage. (*c*) Forces acting on assemblage as shown on plans.

axes of the ring girder and little torsion is introduced (Figure 30-3*b*). In fact, however, the detail shown on the plans had the hopper plates connected to the ring girder at the top [location (2) in Figure 30-3*a*]. Reference to Figure 30-3*c* will indicate that this introduced a pronounced torsional eccentricity. In this case, the calculated stress in the ring girder, due to primary bending plus torsion was 46 kips/in² vs. 24 kips/in² allowable. An unacceptable condition had resulted.

A secondary problem also developed. The plates of the hopper bottom (as is usual) were highly stressed in direct tension. Connection to the ring girder required full penetration welds to develop the strength of the hopper plates. The hopper plates were so thick that they had to be beveled to accomplish full penetration of the welds.[1] The plates, as installed, had not been beveled. Only partial penetration of the welds could be assured. This condition, too, was unacceptable.

The remedial measure consisted of relocating the point of connection of the hopper plates and adding reinforcing plates, as shown in Figure 30-4.

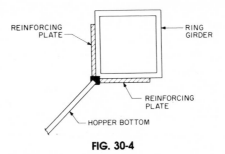

FIG. 30-4

[1] No weld symbols were shown on the drawings.

LEGAL ANALYSIS

There is a substantial distinction to be noted in this case between the technical errors of judgment on the part of the engineer and the consultant as compared to the practical and business errors of judgment which they committed professionally. The testimony of the consultant appeared to satisfy the trial judge that the standard of care necessary to properly design this structure in accordance with the AISC Manual was not met by the consultant. By his testimony, he acknowledged having failed to consider torsional stresses when preparing his calculations. Although a disparity between his drawings and his calculations was pointed out by the engineer, the consultant deemed it unnecessary to undertake a complete review and analysis of his work product in order to clarify the reason for the differentiation pointed out by the engineer.

A different problem confronted the engineer. The senior designer on the project deemed it appropriate to have a nonlicensed engineer perform the review of the calculations and drawings submitted by the consultant. This reliance, without a more experienced judgment participating in this review capacity, left to chance the possibility that a serious design error might go unheeded. It is understandable that time pressures and other factors can result in the delegation of such duties to junior members of an engineer's staff.

It is unrealistic to expect that a senior engineer in all such instances will be undertaking review of drawings and calculations where they are submitted for review. It is not, however, unrealistic to expect that somewhere in the design process, and prior to the final approval stage, such plans and/or calculations will be reviewed for more than mere compliance with the overall design concept for the project. Engineers will have to determine for themselves the necessary degree to which this categorical imperative is applied.

Although many engineers would argue that the review process need not be exhaustive in every situation, certain situations would appear to automatically warrant closer review by an experienced engineer. One such instance is where the design concept is changed during the course of the project. We have already seen a similar situation which occurred in Case 6. Clearly, when there is a shift away from the concept which was the basis for the original design, a total rethinking of all aspects of the project which bear upon that design must likewise be reviewed. The inexperienced drafter or initiate may not have the breadth of experience to envision every manner in which such a change will require adjustment and revisions to the entire work product. Placing this responsibility in the hands of an inexperienced member of the staff greatly increases the chances for an error to pass through the engineer's office undetected.

An entirely separate area of concern in this case involved the engineer's failure to distinguish between the need to review shop drawings for compliance with the overall design concept and the need for a detailed review of certain specifications such as those submitted by a consultant. To a great extent, the engineer becomes the agent of the owner with regard to ensuring that the design fully complies with requisite standards of care which fall within his or her purview. Once vested with responsibility to approve the detailed design, the engineer must clearly understand the scope of his or her work and take all steps necessary to assure attention to such details.

Before the project commences, the engineer must fully understand the nature of the contractual obligation that has been undertaken and the different situations that might be encountered which fall within the scope of required services. Far too often, engineers and architects are found to be using outdated contract provisions which have been extracted from other agreements unsuited to the project at hand. By resorting to such standardization, they are blindly opening themselves to exposure they never intended to assume. Perhaps in the years ahead, the design profession will awaken to the need to have their contracts for services reviewed by knowledgeable counsel. Certainly, contractors and subcontractors have, for years, been employing either their own counsel or that of the sureties who issue performance and payment bonds, to see that their contract documents are in order. We find more frequently than ever that contractors are also utilizing the assistance of attorneys during the course of projects to protect their rights in the event of delays or other problems which may develop.

As a whole, the design profession has woefully failed to recognize the need for competent counseling at every step of the way; certainly the preparation of a contract for projects involving millions of dollars should warrant the careful review and attention of an attorney prior to execution. As this practice becomes more widespread, the incidence of having to defend an architect or engineer in the face of a contract provision which was neither understood nor intended, will become less frequent.

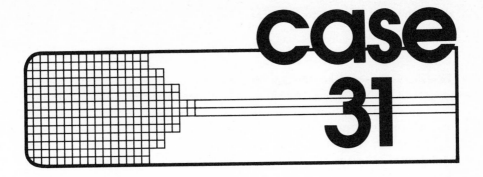

case
31

TYPE OF FACILITY	Multistory apartment building
TYPE OF PROBLEM	Spalling of precast concrete, exterior wall panels

Significant Factors

A. Inadequate provision for movement in connection of panels to structural frame
B. Method of erecting panels
C. Differential thermal expansion
D. Quality of epoxy bonding compound

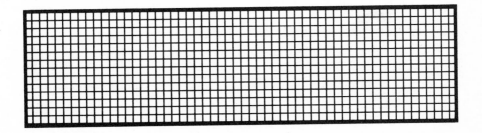

NARRATIVE

This matter arises out of the construction of a senior citizens' rental housing project. The building in question was eighteen stories and was funded through a federal grant. The project was constructed under the auspices of a nonprofit corporation.

The design, as conceived by the architect and his structural engineer, called for a concrete frame. The walls were to be infill, precast concrete panels. The exterior of the panels was to be faced with marble tiles attached by epoxy. The tiles were applied to the concrete panels before they were placed upon the building.

Approximately 7 years after completion of the building, the marble tiles began to fall and continued to do so for a period of 18 months, until repair work was undertaken. Further, cracking of the panels was noted, as well as the resulting penetration of water through these cracks into the interior of the building.

Accordingly, a suit was commenced by the owner against the architect, the contractor, and the manufacturer of the epoxy bonding agent and adhesive. Third-party actions were commenced subsequently by the manufacturer of the epoxy against the subcontractor who fabricated the precast panels under a subcontract which also required him to affix the marble tiles to the panels with the epoxy. The contractor also filed a similar third-party complaint against its subcontractor asserting that its agreement with the subcontractor included a hold-harmless provision. The contractor also joined a subcontractor responsible for all caulking work. Finally, the architect filed a third-party action against the structural engineer.

Damages of $175,000 plus interest were alleged against all defendants. These damages related to repair work, lost income, and outside expert services with respect to the remedial work required.

The owner's position with regard to proving its claim did not set forth a specific theory as to the cause of the observed distress, i.e., the falling tiles, the cracked panels, and the leaks. The owner merely asserted that he contracted with the architect and the contractor for a usable building and that any failures in this regard constituted a breach of contract.

The following theories were advanced by the various parties as possible explanations for the difficulties:

1. The epoxy used was inappropriate or defective.
2. The concrete for the panels was not permitted to cure, resulting in shrinkage.
3. Expansion of the marble and/or incompatibility with the expansion of the concrete.
4. Induced stress on the panels during erection.

5. The use of improper caulking and/or improper application of the caulking.

6. Failure to use appropriate backing for the caulking.

7. The use of an improper method for the attachment of the panels to the concrete columns.

A potential liability of the structural engineer related to his preparation and/or approval of the structural drawings relating to the method of attachment of the precast panels to the concrete columns and floor. Separate drawings were prepared for this detail by both the architect and the structural engineer. Although there were differences in the details, an agreement was reached whereby bidding for the manufacture and erection of the panels would be based on the engineer's design.

After the contract for construction was let, modifications were requested by the subcontractor, who submitted shop drawings. The engineer approved the shop drawings after certain changes which he requested were incorporated. Basically, the change in design provided for an attachment by welding angle irons to metal pieces embedded in the concrete and in the precast panels. Initially, the design called for the attachment to be achieved by bolts, rather than welding. The change in design resulted in less flexibility in the movement of the panels as compared with the use of the bolts; also, the angle iron embedded in the precast panels had the potential for exposure to the elements. (In the final drawing, the angle iron was shown to be embedded in the precast panels only ½ in from exposure to the outside.)

At the deposition of the structural engineer, he testified that although he preferred his original design because of the flexibility of the bolted connection as compared to welding, the final drawing represented an adequate and proper means of attachment. He pointed out that his only real concern was the structural integrity of the concrete members; any questions with regard to specifications for the caulking or the method used in providing backing for that caulking were those of the architect. He also pointed out that the only written specifications which he had drafted related to concrete and that specifications for the precast panels had been prepared by the architect.

As discovery proceeded, it became clear that two separate areas of damage were involved, the first being the tile damage, which amounted to approximately $50,000, and the second being the water infiltration damage, which was approximately $100,000. The parties generally agreed that the structural engineer and the caulking subcontractor had no involvement in the damage to the tile and thus would not be required to make any contribution toward that aspect of the claim. There was, however, agreement that possible causes for the water infiltration damage did involve all of the defendants.

As result of negotiations, a settlement was ultimately reached, where-
upon the separate areas of liability were apportioned among the defendants
and a settlement in the sum of $140,000 was achieved along the following
lines: the architect contributed $42,000, the contractor, $25,000, the manufacturer of the epoxy, $13,000, the fabricator of the wall panels, $40,000,
the caulking subcontractor, $2000, and the structural engineer, $18,000.

TECHNICAL ANALYSIS

Typical details of the subject panels and of their connection to the concrete
frame of the building are shown in Figure 31-1.

Various theories which were advanced to account for the observed delamination of the marble tiles and the water penetration included the following:

1. *Inappropriate or Defective Epoxy or Epoxy Used after Expiration of 6-Month
   Shelf Life*   The details shown in Figure 31-1 had been used, at least
   in similar form, on three previous projects in the United States, and
   on several other previous projects in Europe. No problems had been
   encountered on those previous projects. Further, the epoxy bonding
   material had been tested before manufacture of the panels by casting

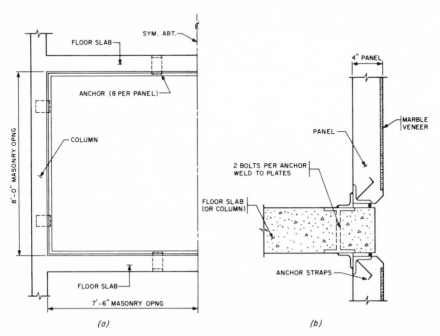

FIG. 31-1   (*a*) Elevation of precast panel. (*b*) Detail of anchor.

full-size samples and trying to pull the veneer tiles loose. The adherence of the veneer in these tests had been satisfactory. The epoxy bonding material itself was a well-known product of a well-known and reputable manufacturer.

2. *Concrete Panels Had Not Been Adequately Cured before Attaching the Veneer* Differential shrinkage would have caused delamination of the veneer. No data in regard to the validity of this contention was found in the files of this case.

3. *Differential Expansion between Marble and Concrete due to Different Coefficients of Thermal Expansion* The effects would be similar to those caused by differential shrinkage. Again, no data in regard to the validity of this contention was found in the files of this case other than the assessment that the dimensions of the panels are so small that calculated total expansion or contraction would be only $\frac{1}{20}$ in, which strain is not enough to cause the observed cracking. Further, if this were the cause, cracking should occur at all connectors, not just around the two at the bottom.

4. *Stress Induced in Panels during Erection* The method used to erect the panels is illustrated in Figure 31-2. The panel was set on the lower floor and the lower anchors were welded together to secure the panel. The panel then was tilted plumb and the top and side anchors welded. This created bending stresses in the lower anchors and in the "lip"

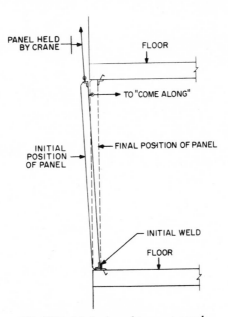

FIG. 31-2 Method used to erect panels.

of the panels in the vicinity of these anchors. The theory was never proved but had to be given some credence, as these extracts from an inspection report indicate:

> At the edge of the panels cracks have occurred in the precast concrete between ½" and 1½" from the outer edge. These cracks generally occur on the outside face of panels opposite points of attachment (at the base of the panels) between the precast panel and the structural concrete.
>
> A water test on the southwest corner at the first floor has shown that water enters the building through these cracks when all other possible penetration routes are sealed. These cracks in the precast concrete appear to be the source of the water penetration problem.

5. *Improper Caulking or Failure to Use Backing for Caulking* The following extract from the above-cited inspection report disputes this contention.

> Caulking performed under original contract to seal joint between precast concrete panel and structural concrete frame appears to be performing its intended function.

6. *Improper Detailing of Anchors for Panels* This discussion centered on the need to provide for relative movement between the panels and the structural frame. The detail shown in Figure 31-1 does not provide for such movement. The original detail had provided for bolted, slotted hole connections in the angles of the anchors. These were eliminated in the "as built" details.

If nothing else, the above listing is a good "checklist" to be considered by any architect or engineer contemplating the use of precast, laminated panels, or, indeed, precast panels in general.

By way of remedial repair work, the marble veneer was fastened to the precast panels by mechanical attachment and the leaking cracks were sealed with an appropriate sealant.

LEGAL ANALYSIS

The significant legal aspect of this case concerns the manner in which the settlement was structured. It was not disputed that there were sufficient questions of fact concerning the contributory causation on the part of each of the defendants for the damages incurred. While it clearly could be asserted that the structural engineer had no involvement with the tile damage, no such assertion could be made with respect to the water damage claim. Similarly, the epoxy manufacturer appeared to face the largest exposure in regard to the tile claim, but there were no grounds for any liability on its part with respect to the water damage claim. The caulking subcontractor also was deemed by the parties to face limited exposure in regard to the water infiltration claim.

By structuring negotiations for settlement along lines which track the exposure for defined categories of damage, it is often possible to attain a meeting of the minds more readily than by seeking a total contribution to a settlement package. This is because parties will more readily admit to a portion of exposure and counsel will be able to convince their clients more readily if the claim is broken down into defined elements.

Some comment should also be made regarding the desirability of using the offices of the trial judge to assist in settlement. Some judges are opposed to participating in settlement negotiations until the time of trial. Others see their participation as a way of clearing congested court dockets and avoiding further pretrial activities. Obviously, different philosophies prevail among the judges in every jurisdiction and it will be incumbent upon the attorneys who seek a negotiated settlement to determine whether an approach to a judge will be fruitful under given circumstances.

In short, the various approaches toward resolving any case by negotiation is limited only by the creativity of the attorneys involved. The structuring of a settlement will generally be most amenable to the parties where it is fair, awards the plaintiff a reasonable percentage of provable damages, and leaves each party to believe that it has achieved a reasonably attainable goal as a result of the negotiation. This does not mean that one or more parties may not successfully conclude negotiations achieving highly favorable results; more than likely negotiations will result in the parties believing that they fell slightly short of achieving a highly desirable result.

case 32

TYPE OF FACILITY Concrete, prestress, posttensioned parking garage

TYPE OF PROBLEM Insufficiency in design of roof under plaza

Significant Factors

A. Failure to consider punching shear at heads of columns

B. Incorrect estimation of loads

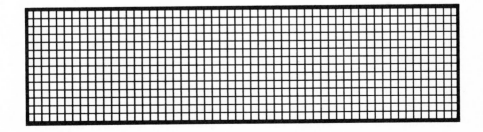

NARRATIVE

This case involves the design of concrete columns supporting a parking garage constructed as part of a large municipal center. The garage had been in use for several months when an inspection disclosed concrete flaking in the vicinity of certain columns. Further investigation disclosed cracks at the top of these same columns. When tapped, several of the columns sounded hollow. The situation was promptly brought to the attention of the architect and structural engineer.

The ceiling of the parking garage was constructed with a 14-in drop panel slab thickened with an additional 6 in at the columns. The slab, of posttensioned construction, supported a landscape plaza covered with earth and planted with trees and grass. In some places the earth was 4 to 11 ft deep.

Promptly after the situation was observed, experts retained separately by the architect and engineer commenced an analysis of the problem. Initially, the problem appeared to be limited to only six columns. However, further review and analysis by the expert for the structural engineer resulted in the conclusion that several, deep-seated problems might exist, not only with the cracked columns, but with shear capacity at the heads of many other columns and with the flexural strength of the slab. The structural adequacy of the roof was questionable. As a result of this indication of serious structural problems, numerous meetings were held between the architect and engineer designed to reach agreement in regard to the repair work which had to be undertaken. The architect agreed to advise the municipality that the portion of the garage which was potentially dangerous should be closed off. Additionally, a plan was implemented to (1) design reinforcement for those columns exhibiting distress, (2) have the independent experts review the several apparent problems to determine the actual strength and, hence, the degree of distress and the need to repair, (3) initiate a scale-model test program to assist in this determination, and (4) use this data to determine whether or not it would be necessary to include additional reinforcement to areas other than the columns showing distress.

From the standpoint of the architect, the entire problem rested on the shoulders of the structural engineer. He asserted that the engineer had failed to properly provide for or consider the effect of the loads on the structure due to the 4 to 11 ft of earth covering the roof of the garage. The early stages of investigation by the experts and attorneys representing the engineer were designed to determine whether other factors may have contributed to this serious condition. These factors included whether or not the concrete was properly poured by the contractor and whether the design criteria were changed by the architect during the project so as to change the effect of the loads on the structure.

At subsequent meetings, the municipality agreed to have the model test performed and discussions, which included the contractor, avoided any mention about which party or parties would be responsible for payment for the testing and subsequent repairs. As a result of these meetings, the structural engineer also agreed to prepare working drawings to obtain bids for any remedial work which might be required.

Subsequently, the model tests were performed, and a detailed stress analysis made of the structure. These tests and calculations disclosed that the presumed distress in the structure was nowhere near as serious as first appeared and that repairs could be limited to reinforcement of a few columns.

Estimates of the repair costs were projected at between $300,000 and $400,000. This did not include a claim by the municipality for loss of revenue from those portions of the garage which were closed off during the repair period or the subsequent claim by the architect for the fees of an independent consultant.

Ultimately, negotiations resulted in an agreement whereby the structural engineer agreed to pay $312,000 for the repair costs. The architect agreed to withdraw his claim against the engineer for out-of-pocket fees to his expert and the municipality agreed to accept a payment of $1000 for its claim of lost parking revenue.

TECHNICAL ANALYSIS

The structure in this case is an underground parking garage (two levels) below the plaza of a major municipal building. The construction is concrete, flat-slab, posttensioned, with drop panels.

The garage had been in use for about 2 to 3 months when a field inspector of the owner noticed "flaking" of the concrete at the tops of several of the columns under the roof level (supporting the plaza and plaza fill). The tops of the columns were cracked and bulged. Tapping gave a hollow sound in the bulged area. Chipping with hand tools showed that about a 2- to 2½-in depth (i.e., down to the spiral reinforcement) of loose concrete could be dislodged from one face of the columns. He called the condition to the attention of the architect. The subsequent investigation revealed the following:

1. The design of the roof had provided for a depth of fill of 4 ft. Actual depth of fill was 8 to 9 ft, average, with a maximum of 11 ft. Calculated bending stresses in the roof slab under the actual loads were excessive, and the columns correspondingly overloaded. This condition existed, generally, throughout the structure.

2. The flaking occurred at the columns along an expansion joint (see Figure 32-1) and was accompanied by cracking, as shown. The designer had

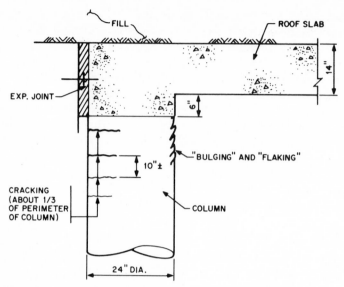

FIG. 32-1 Schematic of column along expansion joint. (Note that cracks are wider at top of column, becoming narrower at lower levels.)

assumed a pinned condition at the tops of the columns. Instead, a significant stiffness existed and a correspondingly significant moment. The cracking and flaking were the result of overstress due to this moment.

3. A critical condition of excessive punching shear stress also existed at the columns along the expansion joint. Investigation suggested that the designer had not checked this area for punching shear.

The investigators concluded that the entire structure was in a dangerous condition and could collapse. In answer to the obvious point that the structure was standing and in full use, the investigators indicated that while this was true, the reserve of strength was only 5 to 10 percent, and that relaxation of the prestress, the development of in-plane tensions due to shrinkage, and degradation due to aging and the environment could produce collapse, if not immediately, in a few years. Much soul-searching by all parties ensued, regarding whether or not to close the structure.

As is usual in cases where a deficiency of design or construction is discovered, every effort was made in the analysis and investigation to discover a sufficient reserve of strength in the structure to justify maintaining it in service and to minimize the need for repair. A number of the points cited below would be of interest to anyone involved in such investigations.

1. The design strength of the concrete was 4000 lb/in². Measurements showed in-place strength of 5000 to 7000 lb/in². The higher strength meant greater resistance to punching shear and some increase in flexural capacity.

2. Design procedures for flat slabs differentiate the column and middle strips, with a greater portion of the panel moments in the column strips. Ultimate strength analysis (collapse slab, for example) treats the full width of the panel as a unit. Inspection of the structure indicated that while calculations showed the column strips to be grossly overstressed, there was no physical evidence of distress. A sophisticated analysis using a computer and a grid model confirmed that the design procedures in ACI 318 are, as expected, conservative and that design for full interaction of the whole panel, i.e., yield of the total section, was justified.

3. As regards the problem of punching shear, differentiation was made between the resistance of a prestressed (posttensioned) structure and one which is not prestressed. For a nonprestressed structure, the investigators indicated that full-scale load tests gave failure loads in good agreement with the predictions of the design procedures in ACI 318, but that prestressing the slab in the vicinity of the column increases the punching shear capacity. The reason indicated was that prestressing counters the in-place tensions in the slab due to 1) diagonal tension, 2) shrinkage, and 3) other sources of reduction in volume of the slab. Such in-plane tensions would, of course, reduce the capacity of the concrete to resist shear at the column head.

4. The design had not considered the secondary moments induced in the slab due to posttensioning. The effects of such moments are illustrated in Figure 32-2. In general, the moments at the supports (negative moments) are reduced and the midspan moments are increased. Recognition of this effect in the building code was first introduced in the 1975 edition of ACI-318.

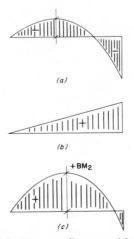

FIG. 32-2 Moment diagrams. (a) Due to gravity loads. (b) Due to posttensioning. (c) Algebraic sum of gravity and posttensioning moments.

5. Load tests, of course, were proposed as the ultimate indicator of the capacity of the structure as built. Based on calculation, however, the assessed danger of collapse of the structure under the test load was such as to preclude using such an approach. Instead, a scale model was constructed and tested; the test indicated an available reserve of strength.

6. A portion of the roof of the garage underlay a street. The applicable building code set forth a requirement to design for a surcharge of 250 lb/ft² to allow for truck loads. Analysis of distribution of wheel loads through the fill overlying the roof indicated sufficient spread of the load to justify use of a lesser surcharge of 140 lb/ft².

In the end, a number of the columns were reinforced. In the case of the columns at the expansion joints (Figure 32-3), the purpose was to reduce the punching shear stresses and to strengthen the columns against the observed flexural distress. Interior columns also were strengthened, in order to stiffen them and thereby reduce the positive moments in the

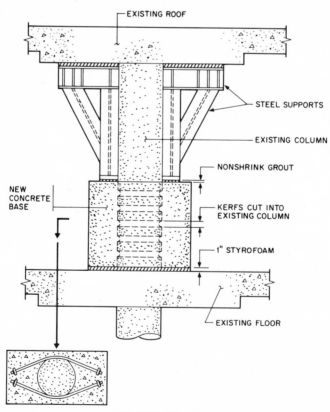

FIG. 32-3 Elevation.

slab (which calculations indicated to be the critical condition). A new wall was added in the portion of the garage under the street, to serve as additional support for the roof.

LEGAL ANALYSIS

In view of the symbiotic relationship between architects and their engineering consultants, a unity of interest usually exists between the two when they are confronted with a design defect. Invariably the best course of action is to achieve an agreed-upon program for corrective measures so that the interest of the owner will at all times be protected. However, not every situation lends itself to such a united front, and at times, the relationship may become adversarial in nature.

In this case, the architect and the structural engineer clearly saw the problem from different vantage points. To his way of thinking, the architect saw a need for immediate remedial repair to ensure that all problems, present and future, would be reviewed and cured at one time. To the trained eye of the engineer, a question existed whether the problem was solely his responsibility, or whether some factor, such as design load criteria, had been altered without his knowledge. Additionally, the experts retained by both the architect and the engineer provided varying preliminary opinions regarding the severity of the problem and this, in turn, resulted in a further breach in the relationship.

Eventually, the mutual respect and professionalism of the design team resulted in an agreed-upon plan of action. This did leave open the question of who would assume the cost of repairs. For purposes of negotiation, counsel for the structural engineer was prepared to admit liability for those damages directly related to the structural problem. This was, however, to exclude those problems attendant to the waterproofing claim. The parties did agree that in the event litigation developed, the municipality would have been required to establish the liability of the responsible party.

Once the load testing was completed, confirming that the original design was adequate if the plaza loading criteria had been kept to a depth of 8 ft, a new element was introduced into the question of liability. A review of the plans in the possession of the engineer showed certain architectural drawings indicating a dirt loading to a depth of 8 ft. However, at some time, a change from 8-ft loading to 11-ft loading was made. Consequently, a question was raised whether the structural engineer was aware of this change and whether or not a failure of the architect to give him timely notification resulted in the punching shear problem.

A legal analysis of the claim by the municipality for loss of revenue was undertaken. For compensable damages, it has been generally held that loss of rentals due to delays occasioned in occupying premises as a result of defects in plans and specifications is too remote and speculative to be considered as an element of damage. In order to recover damages, it is necessary to establish actual monetary damages with reasonable certainty. The documents furnished by the municipality in support of the claim for loss of revenue failed to meet these qualifications and a legal memorandum supporting this point of view was furnished to its counsel and ultimately accepted. An analysis of the criteria used by the municipality to determine the damages also was undertaken and resulted in a conclusion that this claim could not be supported by the documentation produced.

Finally, there was the adamant position of the architect that he was entitled to recover the cost of his expert witness from the structural engineer.

In most jurisdictions, it is accepted law that a party cannot recover damages arising from avoidable consequences. It was argued by the engineer's attorneys that the expert fees of the architect merely involved review of another design professional's work product. Since this review did not entail the preparation of remedial plans and specifications, counsel for the engineer argued that these services were unnecessary. Hence, the position of the engineer was that the architect's experts were retained for the protection of the architect and not as a natural consequence of the alleged malpractice of the structural engineer. Accordingly, their fees did not constitute an item of damage for which the architect could seek recovery against the engineer.

Generally speaking, the measure of damages for negligence in the preparation of design drawings is the cost of repair, where the cost of such repair is not disproportionate to the value of the project. However, in computing the cost of remedial work, one must consider the doctrine of unjust enrichment, or as it is called in design professional cases, the doctrine of beneficial first cost. In its simplest terms, the cost of construction is to be borne by the owner. Should the value of the structure, as constructed, be reduced because of an error in design, the repair cost for this design will be attributable to the party responsible for the defect. The damage which will be assessed is limited to the amount of repair which is necessary to construct a building in accordance with the design approved and accepted by the owner. Such damages do not include any betterment, i.e., improvement to the building or additional to meet upgraded design criteria which may have gone into effect subsequent to completion of the design.

A simple example of this would be the situation in which an architect designs a residence to include a fireplace but makes no provision for a smokestack on top of the roof. As the structure nears completion, it will become apparent that the smokestack has been omitted and, of course, will have to be included before the job can be completed. If the cost of materials and labor for adding the chimney to the plans at this later date will be greater than the cost that would have been incurred by the owner had the design included the chimney at the outset, the architect will be responsible for this additional cost. However, the original cost which the owner would have had to bear had the item been included in the original design, is not a proper element of damage to be assessed against the designer.

Index

About the Authors

BARRY B. LePATNER has had extensive legal experience in architectural and engineering matters. He is active in various sections of the American and New York State Bar Associations.

He has developed a specific expertise in representing design professionals in malpractice actions throughout the United States. He is the author of several articles in this field and has also made various presentations on the subject of professional liability insurance.

He currently practices law in New York City, where he is counsel to architectural and engineering firms. During the year he publishes the *LePatner Report*, a bimonthly newsletter which reports on matters of current import and interest to the design and construction industry.

SIDNEY M. JOHNSON is a graduate of Yale University, where he studied under famed Hardy Cross and where he received both Bachelor's and Master's degrees in Civil Engineering. For over 30 years, Mr. Johnson has specialized in the design of structures and foundations and in problems related to the deterioration of structures and foundations—worldwide—on well over a thousand projects. This broad experience has resulted in the authoring of numerous articles and papers, plus two books: (1) *Deterioration, Maintenance and Repair of Structures* (McGraw-Hill Book Company) and (2) *Design of Foundations for Buildings* (McGraw-Hill Book Company). The latter, written with T. C. Kavanagh, was based on the author's experiences in writing the structural and foundations provisions of the Building Code of the City of New York.

In 1978, Mr. Johnson received the Grand Conceptor Award of the American Association of Consulting Engineers for his work on the development of the Jari River project in Brazil. Mr. Johnson, currently, is a principal in the firm of Berkowitz-Johnson, Consulting Engineers, with offices in Union, New Jersey, and New York City.